T0226663

Reproductive Allocation
in Plants

This is a volume in the

PHYSIOLOGICAL ECOLOGY Series
Edited by Harold A. Mooney
Stanford University, Stanford, California

A complete list of books in this series appears at the end of the volume.

Reproductive Allocation in Plants

Edited by

Edward G. Reekie
Biology Department
Acadia University
Wolfville, NS, Canada

Fakhri A. Bazzaz
Department of Organismic and Evolutionary Biology
Harvard University
Cambridge, MA, USA

AMSTERDAM • BOSTON • HEIDELBERG • LONDON
NEW YORK • OXFORD • PARIS • SAN DIEGO
SAN FRANCISCO • SINGAPORE • SYDNEY • TOKYO
Academic Press is an imprint of Elsevier

Cover photo credit: Eric Kershaw (Kalmia inflorescence)

Elsevier Academic Press
30 Corporate Drive, Suite 400, Burlington, MA 01803, USA
525 B Street, Suite 1900, San Diego, California 92101-4495, USA
84 Theobald's Road, London WC1X 8RR, UK

This book is printed on acid-free paper.

Library of Congress Cataloging-in-Publication Data
Application submitted

British Library Cataloguing in Publication Data
A catalogue record for this book is available from the British Library

ISBN 13: 978-0-12-088386-8
ISBN 10: 0-12-088386-4

For all information on all Elsevier Academic Press publications
visit our Web site at www.books.elsevier.com

Printed and bound by CPI Group (UK) Ltd, Croydon, CR0 4YY
Transferred to digital print 2012

**Working together to grow
libraries in developing countries**

www.elsevier.com | www.bookaid.org | www.sabre.org

ELSEVIER   BOOK AID
            International   Sabre Foundation

Contents

Contributors   ix

1. The Resource Economy of Plant Reproduction   1
P. Staffan Karlsson, Marcos Méndez

    I. Introduction   1
   II. Historical Prelude   1
  III. The Principle of Allocation   3
  IV. Reproductive Effort   3
     A. Definitions   3
   V. Problems in Determining Reproductive Allocation   5
     A. The Currency   6
     B. Definition of Reproductive versus Nonreproductive Plant Parts   7
     C. When Should Reproductive Allocation be Measured?   8
  VI. Dynamic Resource Allocation   8
 VII. Empirical Patterns in Reproductive Allocation   9
     A. RA and Life History   10
     B. RA in Relation to Succession, Competition, and Disturbance   13
     C. RA in Relation to Environmental Stress   18
     D. Genetic Variation in RA   19
     E. What does the Evidence Say?   20
 VIII. Costs of Reproduction   20
     A. Methodological Issues   21
     B. Quantitative Links between Reproductive Allocation and Costs   22
  IX. Conclusions   29
     Acknowledgments   30
     References   30
     Appendix   40

2. Meristem Allocation as a Means of Assessing Reproductive Allocation   50
Kari Lehtilä, Annika Sundås Larsson

    I. Introduction   50
   II. Developmental and Physiological Background of Meristem Allocation   51
  III. Meristem Structure and Generation of Plant Architecture   52
  IV. Axillary Bud Formation and Subsequent Development of the Bud   54
   V. Genetics and Physiology of the Floral Transition   55
  VI. Meristem Types   56
 VII. Meristem Models   57

VIII. The Assumptions of the Models   61
IX. The Impact of Meristem Allocation on Reproductive Allocation   63
X. Plasticity of Meristem Allocation   64
XI. Major Genes of Meristem Allocation   66
XII. Resource Levels and Meristem Limitation   67
XIII. The Function of Dormant Buds   68
XIV. Meristem Allocation as a Surrogate in Estimation of
  Resource Allocation   69
XV. Conclusions   70
  Acknowledgments   71
  References   71

3. It Never Rains but then it Pours: The Diverse Effects of Water on Flower Integrity and Function   75

Candace Galen

I. Introduction   75
II. The Functional Ecology of Water in the Life of a Flower   76
  A. Water Use by Flowers   76
  B. The Water Cost of Flowers   78
  C. Water as a Regulator of Flower Microclimate   79
  D. Water as a Conduit for Environmental Sources of
    Flower Damage   79
III. Water Relations and the Evolution of Floral Traits   80
  A. Floral Traits as Resource Sinks: The Resource Cost Hypothesis   80
  B. Floral Traits and Water in the Flower Microclimate: Parental
    Environmental Effects   82
  C. Plastic Responses of Floral Traits to Water Availability: Impact
    on Plant–Pollinator Interactions   86
IV. Conclusions   88
  Acknowledgments   89
  References   90

4. The Allometry of Reproductive Allocation   94

Gregory P. Cheplick

I. Introduction   94
II. Definition and Analysis of RA in Relation to Allometry   96
III. Allometry Theory and RA   98
IV. Relation of RA to Relative Fitness   102
V. Allometry of Modules   105
VI. Allometry of RA and Plant Life History   109
VII. Determinants of Allometry   119
VIII. Conclusions   120
  References   121

5. Sex-specific Physiology and its Implications for the Cost of Reproduction

126

Andrea L.Case, Tia-Lynn Ashman

I. Introduction   126
II. Sexual Polymorphisms   128
III. Costs of Reproduction   128
   A. Male Costs   129
   B. Female Costs   130
   C. Common Flowering Costs   130
   D. Demographic Costs   131
IV. Avenues for Mitigating the Cost of Reproduction   133
   A. Photosynthetic Reproductive Organs   133
   B. Increased Vegetative Photosynthesis   135
   C. Increased Resource Uptake and Resource Use Efficiency   135
   D. Reabsorption   136
V. Predictions for Sex-specific Physiology Based on Differential Reproductive Costs   136
   A. Predictions for Females and Males   137
   B. Predictions for Hermaphrodites in Monomorphic Sexual Systems: Cosexuality, Monoecy, and Diphasy   138
   C. Predictions for Hermaphrodites in Dimorphic Sexual Systems: Gynodioecy and Subdioecy   139
VI. Potential Causes of Sex-specific Physiology   140
   A. Physiological Differences Reflect Plastic Responses to Contrasting Reproductive Allocation between Sexes   140
   B. Selection Modifies Physiological Traits after the Separation of the Sexes to Meet Differential Reproductive Costs   141
   C. Physiology Changes as a Correlated Response to Selection on Other Traits (e.g., via Pleiotropy or Linkage)   141
VII. Available Data on Sex-specific Physiology   142
VIII. Recommendations for Future Study   148
   Acknowledgments   148
   References   148
   Appendix   154

6. Time of Flowering, Costs of Reproduction, and Reproductive Output in Annuals

155

Tadaki Hirose, Toshihiko Kinugasa, Yukinori Shitaka

I. Introduction   155
II. Modeling of Reproductive Output   156
III. Timing of Reproduction   158
   A. Effect of Nutrient Availability   158
   B. Effect of Germination Dates   160
   C. Effect of Change in Flowering Time   162

IV. Costs of Reproduction   168
   A. Reproductive Effort and the Relative Somatic Cost   170
   B. Nitrogen Use Efficiency   172
V. Reproductive Nitrogen   176
VI. Conclusions   182
   References   182

7. The Shape of the Trade-off Function between Reproduction and Growth   185

Edward G. Reekie, German Avila-Sakar

I. Introduction   185
II. Methods of Describing the Trade-off Function   187
III. The Shape of the Trade-off Function in *Plantago*   191
IV. Impact of Reproduction on Resource Uptake   192
V. Differences in the Resource Requirements of Vegetative versus
   Reproductive Tissue   195
VI. Effect of Nitrogen versus Light Limitation   201
VII. Effect of Growth Pattern   202
VIII. Conclusions   207
   Acknowledgments   208
   References   208

8. On Size, Fecundity, and Fitness in Competing Plants   211

Lonnie W. Aarssen

I. Introduction   211
II. Defining the Components of Competitive Ability for Between-species
   Plant Competition: Lessons from Within-species Competition   215
III. Predicting Fecundity under Competition   216
IV. Relationships among Plant Traits Affecting Fecundity under
   Competition: Alternative Ways to Compete Intensely While Avoiding
   Competitive Exclusion   220
V. Preliminary Empirical Tests   227
VI. Predicting Winners from Rank Orders in Plant Competition: Lessons
   from Sports Competition   231
VII. Conclusions   237
   Acknowledgments   238
   References   238

Index   **241**

Previous Volumes in Series   **245**

Contributors

Number in parentheses after each name indicates the chapter number for the author's contribution

Lonnie W. Aarssen (8)
Department of Biology
Queen's University
Kingston, ON
K7L 3N6, Canada

Tia-Lynn Ashman (5)
Department of Biological
Sciences
University of Pittsburgh
Pittsburgh, PA 15217, USA

German Avila-Sakar (7)
Department of Biology
Mount Saint Vincent University
Bedford, NS
B3M 2J6, Canada

Andrea L. Case (5)
Department of Biological
Sciences
Duke University
Box No. 90338
Durham, NC 27708, USA

Gregory P. Cheplick (4)
Department of Biology
College of Staten Island
City University of New York
Staten Island, NY 10314, USA

Candace Galen (3)
Division of Biological Sciences
105 Tucker Hall
University of Missouri, Columbia
MO 65211, USA

Tadaki Hirose (6)
Tokyo University of Agriculture
Department of International
Agricultural Development
Sakuragaoka 1-1-1, Setagaya-ku
Tokyo 156-8502, Japan

P. Staffan Karlsson (1)
Department of Plant Ecology
Evolutionary Biology Centre
Uppsala University, Villavägen 14
SE-752 36 Uppsala, Sweden

Toshihiko Kinugasa (6)
Graduate School of Life Sciences
Tohoku University
Sendai 980-8578
Japan

Annika S. Larsson (2)
Department of Physiological Botany
Uppsala University
Villav. 6, S-752 36 Uppsala
Sweden

Kari Lehtilä (2)
Department of Natural Sciences
Södertörn University College
S-14189 Huddinge
Sweden

Marcos Méndez (1)
Department of Botany
Stockholm University
SE-106 91
Stockholm, Sweden
Current address:
Área de Biodiversidad y
Conservación
Universidad Rey Juan Carlos
c/ Tulipán s/n., E-28933
Móstoles, Madrid, Spain

Edward G. Reekie (7)
Biology Department
Acadia University
Wolfville, NS
B4P 2R6
Canada

Yukinori Shitaka (6)
Graduate School of Life Sciences
Tohoku University
Sendai 980-8578
Japan

1

The Resource Economy of Plant Reproduction

P. Staffan Karlsson, Marcos Méndez

I. Introduction

The pattern of plant resource investment in reproduction is expected to have important implications for plant life histories and their evolution. In this chapter we will discuss the main concepts and research trends in resource economical analyses of plant reproduction. In particular, we will focus on four concepts: the principle of allocation, reproductive effort, reproductive allocation, and cost of reproduction. Although these concepts were intended to have general application, we will focus on their use when studying plant reproduction. We will focus on how their meaning and use have varied over time. Our emphasis on concepts seems justified because, despite substantial amount of empirical research, there has been considerable confusion in this research field. This confusion has negatively affected the theoretical, methodological, and experimental progress in this field. We will also present a brief survey of the achievements in this field during the past 30 years and give some suggestions for the future direction.

Since many reviews on plant reproductive biology, with a more or less strong focus on resource economy, have been published in the past two decades (e.g., Antonovics, 1980; Willson, 1983; Bazzaz and Reekie, 1985; Goldman and Willson, 1986; Bazzaz *et al.*, 1987, 2000; Lovett Doust, 1989; Bazzaz and Ackerly, 1992; Reekie, 1997, 1999; Obeso, 2002), we have made no attempt to include all relevant articles. Rather we have focused on seminal papers, reviews, and some examples of empirical studies.

II. Historical Prelude

At least since ancient Greece, there are written records of man's awareness of the fact that organisms vary in reproductive output and that reproduction may have negative consequences for plant performance (Jönsson and

Tuomi, 1994). Arguments relating reproduction to resource economy had been used since the nineteenth century (Mattirolo, 1899, as cited in Bazzaz *et al.*, 1987). However, the foundations of current research on reproductive strategies in relation to the use of resource were first published around 1930 (e.g., Fisher, 1930) when arguments of physiological costs of reproduction were explicitly related to life history phenomena and evolution (Lovett Doust, 1989; Jönsson and Tuomi, 1994). Some later articles on the clutch sizes of birds (Cody, 1966; Williams, 1966a) are commonly referred to as the starting point for more intensive studies of reproductive biology in relation to life history strategies more or less explicitly built on arguments of the use of limited amounts of resources. The large impact of the articles by Cody (1966) and Williams (1966a) was due to two hypotheses, "the principle of allocation" and "reproductive effort" (RE), proposed by them in these articles.

Within plant ecology, one of the first systematic attempts to seek patterns in plant reproductive investments was made by Salisbury (1942), who screened an impressive array of the British flora and recorded their seed sizes and numbers. Harper (1967), in his address to the British Ecological Society, critically evaluated the concepts proposed by Cody (1966) and Williams (1966a,b) by using examples from the plant kingdom. The first empirical article explicitly using arguments in resource (energy) economy in relation to life history strategies in plants was by Harper and Ogden (1970), who presented the dynamics of biomass allocation in *Senecio vulgaris*. Thereafter, a relatively large number of studies on plant reproductive allocation were published (e.g., Solbrig, 1971; Gadgil and Solbrig, 1972; Abrahamson and Gadgil, 1973; Gaines *et al.*, 1974; Ogden, 1974; Hickman, 1975; Hickman and Pitelka, 1975; Pitelka, 1977; Abrahamson, 1979; Soule and Werner, 1981). Initially this work focused on biomass allocation (as a reflection of carbon or energy allocation) and/or the calorific contents of biomass compartments. Allocation of carbon was not studied at this time. From the second half of the 1970s, reproductive allocation of other nutrient resources was also studied (van Andel and Vera, 1977; Lovett Doust, 1980a,b; Abrahamson and Caswell, 1982; Whigham, 1984).

The first empirical studies on reproductive effort in the 1970s were framed within the r–K theory and the definition of plant strategies. Differences in reproductive effort between annuals and perennials and along successional and disturbance gradients were investigated. From these studies, interest in the effect of density and size-dependence of reproductive effort developed during the 1980s. Simultaneously, interest in the responses of plants to different stresses began to include the study of reproductive effort in the 1970s.

At this time also, studies including demographic costs of reproduction started to emerge. The observations of a negative effect of current reproduction on current growth of subsequent fecundity or survival were in fact old and well known in the horticultural and silvicultural literature (reviewed in

Leonard, 1962; Kozlowski, 1971). The first instance, to our knowledge, of a study addressing cost of reproduction in plants within a life-historical perspective was made by Sarukhán (1974).

III. The Principle of Allocation

The ideas underlying the principle of allocation can be traced back to Goethe (see Lovett Doust, 1989). The principle of allocation was first mentioned by Cody (1966) where he cites unpublished work by Levins and MacArthur; "It is possible to think of organisms as having a certain limited amount of time or energy available for expenditure, and of natural selection as that force which operates in the allocation of this time or energy in a way which maximises the contribution of a genotype to following generations." This principle explicitly states that it deals with the allocation of resources that are available in limited amounts. The interpretation of this term has been clear cut and straightforward: it states that resource allocation patterns are adaptive and shaped by natural selection. Based on this principle, investigators have predicted that allocation of resources to one function (such as reproduction) should have negative consequences for other functions (mainly growth and defense). In other words, the resource in focus is limiting plant performance. However, this assumption has rarely been evaluated explicitly. We will come back to this issue later.

IV. Reproductive Effort

A. Definitions

There are two partially intermixed definitions of reproductive effort (RE). One of them refers to a descriptive account of resources invested in reproduction, while the other one refers to the effort or somatic cost that such reproductive investment entails to the organism. This duality of RE's definition until now has led to some problems in terminology and understanding (Tuomi *et al.*, 1983; Bazzaz and Ackerly, 1992).

The interpretation of RE as a cost is defined by Williams (1966a) not in energetic terms, but in terms of demography (effects on survival and fecundity). Later, authors working from a resource economy perspective also adopted this interpretation. Bazzaz and Reekie (1985) summarize it in the following words: "ideally then [...] any measure of the 'effort' involved in reproduction would be based not on the allocation of resources to reproduction but on *the degree to which vegetative processes were decreased by reproduction*" (our italics). The interpretation of RE as a cost is also present in several subsequent reviews (Lovett Doust, 1989; Stearns, 1992).

The descriptive interpretation of RE can be traced back to Fisher (1930), who implicitly refers to reproductive effort as the proportion of resources diverted to reproductive organs (or reproduction). This view of RE underlies the definitions found in many subsequent contributions (Stearns, 1976; Schaffer and Gadgil, 1977; Antonovics, 1980; Evenson, 1983; Willson, 1983; Kozlowski, 1991; Reekie, 1999). For example, Willson (1983) defines RE as "the proportion of the total resource budget of an organism that is devoted to reproduction." At least Willson (1983) is aware of the potential uncoupling between RE and cost, but it is uncertain to what extent other authors have really separated descriptive and cost sides of the definition.

This dichotomy in the meaning of RE is mirrored in the propositions for its empirical measurement. Bazzaz and Reekie (1985) provide several formulae to estimate RE. They are divided into two categories: direct measures and indirect measures of RE. Direct measures of RE agree with the descriptive definition of RE by previous authors (see references above), while indirect measures of RE are, indeed, equivalent to what other authors consider somatic costs of reproduction (RE6 of Bazzaz and Reekie, 1985, is identical to relative somatic cost of Tuomi *et al.*, 1983).

Some authors (e.g., Bazzaz *et al.*, 1987; Marshall and Watson, 1992) use the term "reproductive allocation" (RA) as synonymous with the descriptive interpretation of RE (e.g., Bazzaz and Ackerly, 1992). We have traced this term to Hickman (1975), although it is not regularly found in the literature until the second half of the 1980s. Yet another term is "reproductive investment" (RI), which Tuomi *et al.* (1983) used to mean the *absolute* amount of resources put into reproductive structures but utilized by other authors (e.g., Ashman, 1994) as synonymous with RA. However, in the estimates of RI, Ashman (1994) includes also allocation to pollen and nectar and resorption of nutrients from reproductive support organs before they are abscissed (cf. Dynamic resource allocation). For additional terminology (reproductive index, RI; harvest index, HI), see Aronson *et al.* (1993). Proliferation of terms for the same concept is regrettable, but in this case, RA deserves some consideration. RA is attractive because it lacks the ambiguity of RE. Tuomi *et al.* (1983) explicitly stated the double meaning of RE and made clear the theoretical relationships between RE and cost of reproduction. RE will be a reflection of costs of reproduction, as demanded by the "cost interpretation," only under a specific allocation system where reproductive and nonreproductive individuals have access to the same amount of resources (this problem will be discussed in the chapter).

Bazzaz and Ackerly (1992) use both RA and RE in the same review chapter and clarify the difference between the two: "the concept of reproductive effort, which is often equated with reproductive allocation (RA), should refer specifically to the individual's net investment of resources in reproduction which is diverted from vegetative activity [...] we discuss several

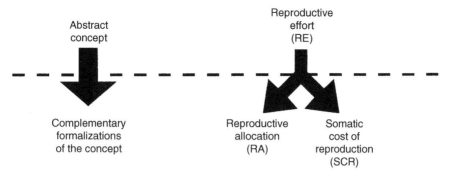

Figure 1.1 Proposed relationships between the terms "reproductive effort," "reproductive allocation," and "somatic cost of reproduction" discussed in the chapter.

factors which decouple RA and effort, and argue that these two must be conceptually distinguished in future research." Thus, these authors make synonymous use of RE and somatic cost of reproduction (SCR).

Despite the attempts of conceptual cleaning by Tuomi *et al.* (1983) and Bazzaz and Ackerly (1992), no consensus seems to have been reached and RE, RA, and cost of reproduction coexist in the recent literature. A way to synthesize these conceptual and semantic issues is to use the distinction made by Ford (2000) between concepts from research and concepts by measurement. We consider RA and SCR as two related, complementary ways (concepts by measurement) to formalize the original, abstract idea (concept from research) of RE (Fig. 1.1). According to this distinction, in the rest of this chapter we will use RA when talking about resources apportioned to reproduction, SCR when describing the consequences of such apportionment for somatic functions, and RE only when talking in abstract terms or mostly when referring to authors who adopted that terminology. The uncoupling between RA and SCR will be addressed below.

V. Problems in Determining Reproductive Allocation

Reviews on plant reproductive strategies (e.g., Antonovics, 1980; Thompson and Stewart, 1981; Bazzaz and Reekie, 1985; Goldman and Willson, 1986; Bazzaz *et al.*, 1987; Lovett Doust, 1989; Reekie, 1999) give a strong impression that this subject has been plagued with methodological problems. In part, these problems relate to the use of the term reproductive effort (see above), but much of them are related to the methods used for determining RA (mostly RE in the phrasing of these authors). In particular, three major areas of difficulties in the measurement of RA in plants have been identified and discussed: (1) the currency problem

(2) how to define reproductive versus vegetative structures, and (3) when to measure reproductive allocation.

A. The Currency

When the principle of allocation was formulated, two kinds of resources were in focus, energy and time. This, in combination with the general interest in energy flow in plant and systems ecology at that time (e.g., Odum, 1961), made energy a natural choice of currency to focus on also when studying plant reproductive allocation.

For plants, different "food" resources (water, light, and 14 mineral nutrients for most plants) are acquired more or less independently, and their availability vary asynchronously in space and over time (cf. Gleeson and Tilman, 1992; Jackson and Caldwell, 1993; Marschner, 1995; Lambers *et al.*, 1998). Despite this, there are few attempts in the reproductive resource allocation literature to evaluate whether or not a particular resource actually is limiting the plant performance. This assumption of the principle of allocation is related to the discussion of the limitation of plant production in the agricultural and horticultural sciences. Since Liebig (1855) formulated the "minimum law," the extent to which plants are limited by one or many resources has been addressed in many studies, particularly regarding plant production and optimization of crop production, but also for natural plant communities (Gleeson and Tilman, 1992; Rubio *et al.*, 2003).

Several authors have argued that plants are balanced systems where allocation is adjusted to obtain a balance between different resources (Bloom *et al.*, 1985; Chapin *et al.*, 1987; Bazzaz, 1997). Using carbon as a common currency, the plant can modify the allocation pattern to achieve a balanced resource acquisition. In the short term, however, availability of various resources can be expected to vary asynchronously; mineral nutrients, water, and light show large temporal asynchronous differences in availability (cf. references above). An analysis of a plant at one point in time may suggest that the plant was poorly adapted when in fact it was maximizing overall growth rate (Gleeson and Tilman, 1992). Experiments evaluating the minimum law versus the multiple limitation hypothesis on growth of *Lemna minor* found that neither hypothesis adequately predicted plant responses (Rubio *et al.*, 2003).

With regard to plant allocation to reproduction, it has been pointed out that reproduction and somatic growth can be limited by different resources (Willson, 1983; Reekie *et al.*, 2002; cf. also Reekie and Avila-Sakar in this volume).

The first studies quantifying reproductive allocation were focused entirely on energy allocation as reflected by biomass or calorific content. Some authors concluded that biomass allocation was a good estimate for energy allocation (Hickman and Pitelka, 1975; Abrahamson and Caswell, 1982) while others came to different conclusions (Funk, 1979 as cited in

Antonovics, 1980; Jurik, 1983; Jolls, 1984). In these studies, energy content of various plant compartments were then compared at one point in time. However, evidence was accumulating from the 1960s that reproductive structures may have a considerable photosynthesis of their own. This was first investigated for the carbon contribution from ear photosynthesis for grain filling in cereals (Watson *et al.*, 1963; Thorne, 1963, 1965; Carr and Wardlaw, 1965; Evans and Rawson, 1970). Later also other species were found also to show similar patterns (Maun, 1974; Ong *et al.*, 1978, Bazzaz and Carlsson, 1979; Bazzaz *et al.*, 1979). Furthermore, it was found that nectar production can consume significant amounts of carbon (Southwick, 1984), and the presence of reproductive structures may enhance leaf photosynthetic capacity (Reekie and Bazzaz, 1987a; Laporte and Delph, 1996). However, in some cases, reproduction has no (Houle, 2001) or a reverse effect (Karlsson, 1994; Obeso *et al.*, 1998) on leaf photosynthesis. Thus, biomass or energy content of reproductive structures at one point in time is not a good measure of the total energy or carbon investment in reproduction. It was thus concluded that energy might not be the most important resource limiting plant performance and that the investments of other resources were likely to be more critical for plant performance (Antonovics, 1980; Lovett Doust, 1980 a,b; Thompson and Stewart, 1981). In fact, Harper (1977, p. 656) stated "… the green plant may indeed be a pathological over-producer of carbohydrates and the resource that needs critical allocation may often be something other than time or energy." In contrast, Bazzaz and Reekie (1985) argued that it is the rank order of reproductive effort that is important, not its absolute value. According to Abrahamson and Caswell (1982) the rank order of reproductive effort in comparison of populations seems independent of the currency used.

Although it may be difficult to find simple universal generalizations on this theme, it seems safe to conclude that productivity is often simultaneously limited by several resources (but see Rubio *et al.*, 2003) even though nitrogen commonly is the resource most strongly limiting production of terrestrial plants (Ågren, 1985). It should also be borne in mind that there may be large deviations from this typical case because the literature addressing plant productivity shows examples of many different response patterns to the manipulation of one or several resources (cf. Gleeson and Tilman, 1992). In theory different currencies are interconvertible, but since conversion coefficients are seldom known, the choice of currency is important (Chapin, 1989).

B. Definition of Reproductive versus Nonreproductive Plant Parts

Plants are very variable in both morphology of reproductive parts and in how clearly these parts are distinguishable from somatic parts. This results in difficulties in finding one definition of reproductive parts that is generally applicable. For example, Gadgil and Solbrig (1972) included all reproductive structures in RE, while Harper and Ogden (1970) included seeds only in RE. Furthermore, how to define "reproductive structures" has been under discussion repeatedly (e.g., Thomson and Stewart, 1981; Bazzaz and Reekie, 1985; Reekie, 1999; Bazzaz *et al.*, 2000). Bazzaz and Reekie (1985) argued that not only reproductive propagules but also male flower parts and other ancillary and support structures must be included. They show further, the dilemma of differentiating between somatic and reproductive parts of many grasses where stem internodes elongate when the plant flowers. This can easily extend into a view where the entire plant can be defined as reproductive; "in one sense all structures and activities are reproductive since the ultimate objective of all plant growth is to produce offspring" (Reekie and Bazzaz, 1987a). Currently there is no definition of reproductive structures that has obtained general acceptance and that works on all plants. The closest to a general consensus that we can see is the definition given by Thompson and Stewart (1981): "... all structures not possessed by the vegetative plant" (p. 208). However, they end their article by stating "we would not at this stage ... like to be too dogmatic about how reproductive structures should be defined" (p. 210).

C. When Should Reproductive Allocation be Measured?

Resource allocation within plants is highly dynamic, some elements or compounds are repeatedly moved between different plant compartments. Particularly when approaching maturity or the abscission of plant parts large changes in resource contents may occur rapidly (Antonovics, 1980; Chapin, 1980). The outcome of a comparison between resource levels in various compartments may thus be strongly dependent on when it is determined. Most commonly, reproductive allocation has been quantified close to seed maturation, this is however not always easy to determine. For example, Harper and Ogden (1970) wrote "there is little difficulty deciding approximately when a small plant with a few seed heads is mature (i.e., when maximum net production has been achieved), but a large plant with many heads at different stages of development presents considerably more difficulty." Furthermore, at the "maturity" stage some resources in reproductive support structures will be resorbed and made available for continued growth or future reproduction (Chapin, 1989; Ashman, 1994). Bazzaz and Ackerly (1992) have distinguished between "standing RA" (a snapshot measure), "short-term RA" (over a short interval, relative to plant lifetime), and "lifetime RA" (over the entire life span of an individual).

VI. Dynamic Resource Allocation

Following the criticism of the "static" methods to measure reproductive allocation at one point in time, the need for better, more "dynamic," methods for assessing energy allocation was obvious. The first attempt to, in detail, measure actual investments in reproduction in terms of carbon was to our knowledge made by Jurik (1983), where carbon dioxide exchange of various reproductive parts was taken into account. A few years later Reekie and Bazzaz (1987a – c) performed a comprehensive analysis of both carbon and nutrient allocation in reproductive and nonreproductive individuals of *Agropyron repens* genotypes grown at different resource availabilities. Their results confirmed earlier conclusions that biomass allocation is not a good measure of nitrogen and phosphorus allocation. However, when taking respiration and photosynthesis into account, energy, nitrogen, and phosphorus allocation matched closely. They also concluded that carbon allocation tends to be biased towards the most limiting resource (Bazzaz and Reekie, 1987b).

An even more detailed analysis was made by Ashman (1994), who in addition to photosynthesis and respiration in reproductive parts, quantified investments in nectar, pollen, and nutrient resorption from reproductive support structures. She found that the dynamic estimates of resource allocation of *Sidalcea oregana* were better predictors of future (next year) reproductive investments than the static estimates. There were, however, no large differences between static and dynamic allocation measures to predict next year's reproductive allocation.

The "dynamic approach" is an important methodological improvement particularly when studying carbon or energy allocation. However, this approach is also relatively demanding and requires sophisticated instrumentation to measure carbon exchange in various, often small, plant parts. It is thus difficult to apply on a larger number of populations or species and few have followed their examples and quantified whole plant carbon dynamics during reproduction (see however, Laporte and Delph, 1996). Some studies have made detailed analyses of carbon assimilation and translocation at the level of individual flowers or inflorescences (Jurik, 1985; Williams *et al.,* 1985; Galen *et al.,* 1993; McDowell *et al.,* 2000; see also references in the "Currency" (Section V.A) and Watson and Casper, 1984, and Obeso, 2002, for reviews on this theme). If the study is restricted to nutrients (Hemborg and Karlsson, 1999; Eckstein and Karlsson, 2001), the analysis of dynamic resource allocation becomes considerably simpler.

VII. Empirical Patterns in Reproductive Allocation

The empirical literature on RA is too vast to be comprehensively covered in this review. In this section, we mostly focus on the classical literature included in previous reviews (Soule and Werner, 1981; Evenson, 1983; Willson, 1983; Fenner, 1985; Bazzaz *et al.,* 1987; Hancock and Pritts, 1987; Marshall and Watson, 1992), on which most of the current ideas about RA have been built. Agreement between authors studying theoretical expectations and empirical findings has been limited. This prompted a review article by Bazzaz and

Reekie (1985), who presented three possible explanations for this: (a) the theories were incorrect or not general enough, (b) there were methodological problems (currency, identification of reproductive structures, and processes) in estimating RA and/or (c) the tests of the theoretical expectations were faulty.

Assessment of point (a) from Bazzaz and Reekie (1985) is beyond the scope of the present review. In a previous section we have addressed the point (b). Here, we revisit the empirical patterns of plant RA in the light of the point (c), i.e., the adequacy of the tests of the theoretical predictions. When relevant, we divide our review of empirical results into intraspecific and interspecific patterns, an important distinction that has not been stressed by all previous authors.

Three kinds of patterns of RA are not dealt with here: division of resources between sexual reproduction and vegetative propagation, effects of herbivory on RA, and patterns of RA in dioecious and gynodioecious plants (for the latter see e.g., Ågren *et al.,* 1999). Furthermore, the relationship between RA and plant size is reviewed in another chapter of this book and is not considered here. However, some reference is made to that literature when it sheds light on other patterns in which lack of size-dependence of RA can introduce confusion.

A. RA and Life History

RA is central to several questions in life history theory, as it is assumed that patterns of RA are directly related to fitness (Calow and Townsend, 1981). In the words of Stearns (1976): "The key life-history traits are brood size, size of young, the age distribution of reproductive effort, the interaction of reproductive effort with adult mortality, and the variation in these traits among an individual's progeny."

Life history theory has produced a number of predictions about the way in which RA should be apportioned along the life of an organism (Roff, 1992). In the 1970s, these predictions were mostly framed in the parlance of the now outdated r – K theory (Solbrig, 1980; Reznick *et al.,* 2002). In the next section, we examine some of those predictions in a more demographic or modern life-historical framework.

1. Responses of RA to Differential Juvenile and Adult Mortality   When adult mortality rate due to environmental causes exceeds juvenile mortality rate, RA should be high (Willson, 1983). Although short adult life expectancy can select for high RA, the converse (that high longevity favors low RA) is not necessarily true (Willson, 1983). If juvenile survivorship is regularly low, there is less benefit to be obtained from reproduction and selection may favor lower RA, but only if that lower RA brings with it either an improvement of juvenile survivorship or a compensatory increase in adult longevity, and probability of future reproduction (Willson, 1983). This prediction derives from the arguments about the evolution of semelparity versus iteroparity (Cole, 1954; Harper, 1967; Charnov and Schaffer, 1973). The consequences of variable and unpredictable survival of young are controversial; models have been formulated which predict low RA and iteroparity, high RA and semelparity, and mixtures of degrees of delayed semelparity.

We are aware of a single data set that allows testing the predictions made earlier at an interspecific level. Karlsson *et al.* (1990) studied RA for three iteroparous species of *Pinguicula.* Svensson *et al.* (1993) studied the demography for the same three species. In agreement with the prediction, *P. villosa,* the species with highest adult mortality and lowest expected longevity, showed the highest RA.

At an intraspecific level, De Ridder and Dhondt (1992a,b) provide RA data for *Drosera intermedia* in three habitats showing contrasting adult and juvenile survival. In a "path" habitat, low adult survival and high RA were found, while in a "seepage" habitat, high adult survival, low juvenile survival, and low RA were found. Predictions were not fulfilled, however, in a "pool edge" habitat, in which both low adult survival and RA were found. De Ridder and Dhondt (1992b) explain this as a consequence of the marginality of this habitat for the species. There, an environmental constraint, namely inundation, inhibits flowering and prevents the attaining of an optimal RA (de Ridder and Dhondt, 1992b).

These two studies provide tentative support for the predictions relating demography and patterns of RA. More studies are needed.

2. RA in Semelparous versus Iteroparous Species   Life history theory predicts that semelparous species should have higher RA than iteroparous species (Roff, 1992). The reason is that if survival is reduced to zero in semelparous species, then there is no point in saving resources for the future and those should instead be diverted towards reproduction. There are some problems concerning causality, which make empirical tests of the hypothesis more difficult. Cockburn (1991) points that higher RA in annuals compared to perennials will not reveal whether the high RA prompted the annual life history (as a resource economy argument would imply) or whether a high RA was a secondary development after annual life history evolved by any other reason (like low survival probability, as assumed by the demographic theory). This is still an open question in the study of plant life histories.

A number of articles during the 1970s and early 1980s tested that prediction for plants (Appendix 1) have been reviewed before (Fig. 1.2) (Harper *et al.,* 1970; Evenson, 1983; Willson, 1983; Bazzaz *et al.,* 1987; Hancock and Pritts, 1987). In general, empirical evidence based on the study of related species supports the prediction (Harper *et al.,* 1970; Willson, 1983; Bazzaz *et al.,* 1987; Hancock and Pritts, 1987). Willson (1983) and Bazzaz *et al.* (1987) also report cases of annuals with low RA and perennials with high RA.

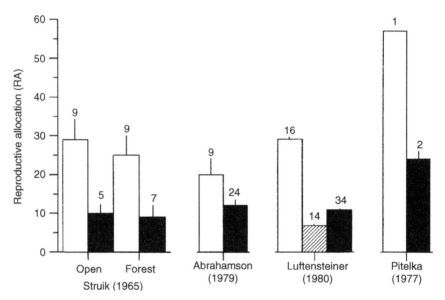

Figure 1.2 Mean value + SE of RA for annuals and perennials in four representative studies. Open bars, annuals; closed bars, perennials. For Luftensteiner (1980), hatched bar, perennial geophytes; closed bar, perennial hemicryptophytes. Number above each bar indicates sample size. For more details, see Appendix 1.

As most of those examples came from studies of single species, it is difficult to put them in context.

Several caveats affect these early tests. Studies by Struik (1965), Abrahamson (1979), and Luftensteiner (1980) were not restricted to related species, and are susceptible to problems of phylogenetic nonindependence (Harvey and Pagel, 1991). In Luftensteiner (1980), "noise" introduced by the use of different methodologies, concepts of RE, measures of RA, and so on could also affect the results. Limitation to interspecific studies comparing closely related (same genus) species should reduce (but not totally eliminate) undesirable phylogenetic effects. Of the studies using related species, only Primack (1979) included a broad range of species, while the others were limited to a comparison between 2–4 species. Only two studies, Gaines *et al.* (1974) and Primack (1979), included some kind of correction for size-dependence of RA (Samson and Werk, 1986; Ohlson, 1988; Weiner, 1988). In two studies (Gaines *et al.*, 1974; Primack, 1979), species from different successional stages were combined and this makes the analyses susceptible to confounding factors of habitat (see below).

Hancock and Pritts (1987) gathered information for Asteraceae, taken from many studies, and found support for the prediction. However, they did not take into account successional stage, which is a confounding factor (discussed subsequently). When the comparison is made separately for "field" annuals versus "field" perennials, there is no significant difference in RA results (mean ± SD for annuals: 23.8 ± 13.0, $n = 4$; perennials: 21.2 ± 11.2, $n = 11$; Mann–Whitney U: 21.0, $p = 0.946$). The analysis of Asteraceae by Hancock and Pritts (1987) is also potentially weakened by the neglect of phylogenetic effects and size-dependence of RA.

Wilson and Thompson (1989) are the only ones who have explicitly controlled for size-dependence of RA. Taking into account maximum height of 23 species of Poaceae, they showed a higher RA for annual versus perennial species. Silvertown and Dodd (1997) have reanalyzed many previous data sets (e.g., Abrahamson, 1979; Hancock and Pritts, 1987) using phylogenetically independent contrasts (PICs). Evidence for the prediction has been found for several, but not all, of them. Joint account of phylogeny and size-dependence of RA has not been done, as far as we know.

3. RA and Age   Life history theory has several predictions about how RA should change with age. On one side, as individuals get older their residual reproductive value is supposed to decrease and their RA should always increase with age (Williams, 1966b; Roff, 1992). On the other hand, senescence theory states that RA should peak early after maturity and then decline (Stearns, 1976; Thomson, 1996). These predictions have rarely been tested for plants, as age cannot be determined for many of them unless long-term demographic studies are carried out.

An increase of RA with age has been shown for the palm *Astrocaryum mexicanum* (Piñero *et al.*, 1982), *Solidago pauciflosculosa* (Pritts and Hancock, 1983), *Drosera closterostigma* (Karlsson and Pate, 1992), and beech *Fagus sylvatica* (Comps *et al.*, 1994). On the other hand, *Vaccinium corymbosum* showed a peak in RA early in life and a subsequent decline (Pritts and Hancock, 1985). The extent to which the patterns documented in these studies are related to changes in plant size with age is unknown.

B. RA in Relation to Succession, Competition, and Disturbance

Harper (1967) proposed a relationship between successional status of plant communities and expected RA, mediated by competition: "colonizing plants should devote more of their resources to reproduction than plants from more mature habitats. Individuals in open environments will face little interference from neighbours; consequently, the chance of leaving descendants will be closely linked to fecundity. In a crowded community where resources are more limiting, however, individual success is much more dependent upon the ability to capture a share of the resources. The individual that sacrifices competitive ability for fecundity may not survive to reproduce." A similar reasoning has also been made by Willson (1983). Harper *et al.* (1970) suggested that the intensity of competition is corre- lated with habitat maturity and, as a result, early successional species are primarily annuals with large RA, while species from mature communities are often perennial and have lower RA.

These ideas were subsequently framed within the predictions of the r–K theory (Gadgil and Solbrig, 1972). Disturbance and competition were seen as ecological factors determining the strength of density-dependent mortality, which was the decisive variable influencing the r or K selection of species (Gadgil and Solbrig, 1972). Disturbance is considered to be higher in early successional habitats when compared to late successional ones (Grime, 1979). Thus, a logic connection was established between habitat maturity, disturbance, and competition. In stable environments, where mortality is usually density-dependent, small RA should be favored. In fluctuating environments, where the usual situation is one of density-independent mortality, large RA should be favored (Stearns, 1976).

Alternative predictions can be derived from other theoretical frames, which incorporate a demographic component (iteroparity versus semelparity) into the effects of competition or disturbance. According to Schaffer (1974), when iteroparity is the optimal strategy in a constant environment, RA should be lower in a variable than in a constant environment (opposite to r–K theory). Conversely, when semelparity is optimal in a constant environment, RA should be higher in a variable than in a constant environment (in agreement with r–K theory). As far as we know, these more sophisticated predictions have never been tested in plants. In the following sections, we will review the empirical evidence concerning the patterns of RA along successional, competition, or disturbance gradients.

1. RA and Succession   Empirical evidence has been reviewed by Soule and Werner (1981) and by Hancock and Pritts (1987) in a vote count way. Only two of seven articles reviewed by Soule and Werner (1981) showed a clear fit to the predictions of Harper (1967) of decreasing RA with increasing successional maturity of the habitat. Hancock and Pritts (1987, and see Appendix 2 for additions, corrections, and exclusions) found a mixed evidence, potentially dependent on the level (intraspecific, genus, community) studied. For populations of the same species (26 species, 15 articles), 50% of the studies showed the predicted trend. For species of the same genus, studies (four genera, six articles) fulfilled the prediction in all but two cases (Fig. 1.3). Five papers addressed the question at the community level and one of them fitted the hypothesis (Fig. 1.3).

Some caveats of these data sets need to be considered. First, the theory is not explicit about the shape of the response to the successional gradient, and this made it difficult to decide whether some empirical patterns fitted the prediction or not. For example, some studies recorded responses with an outlier (Roos and Quinn, 1977), or the responses of RA to the successional gradient were asymptotic rather than linear (Hickman, 1977; Newell and Tramer, 1978; Soule and Werner, 1981).

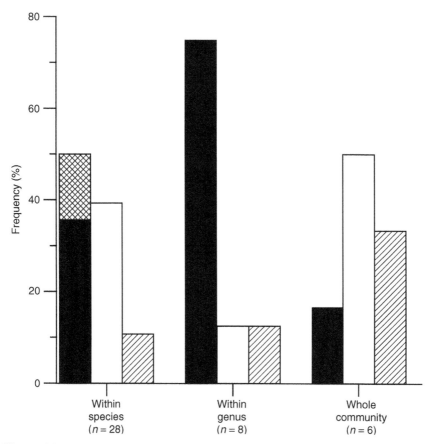

Figure 1.3 Percentage of studies showing higher RA in earlier successional habitats (closed bar), no such trend (open bar), or uncertain pattern (hatched bar), according to the level of comparison (within a single species, species of a same genus, or whole community). The double-hatched bar indicates studies showing the right trend, but not statistically significant.

This difficulty of assessment is related to the number of populations compared. Usually, only 2 – 4 populations or points along a successional gradient have been compared and temporal or spatial replication within those "points" has rarely been included (Appendix 2). As a consequence, populations departing from the expected trend cannot be properly assessed. Lack of replication makes it impossible to assess whether those departures are due to peculiarities of populations unrelated to their successional status, or to a real departure from the predictions. Soule and Werner (1981) have stressed the importance of both spatial and temporal replication due to phenotypic plasticity in RA.

Another caveat which makes it difficult to interpret the results is the nature and extent of the gradients studied. Only a few studies (Appendix 2) provide time since last disturbance for all points of the gradient. More often successional status has been assessed as degree of cover, vegetation composition, or other physiographic traits (Appendix 2). This can be a safe procedure when a complete gradient is studied, e.g., open fields to forest. However, in several studies, a shorter gradient was covered (Appendix 2) and in some cases we could not identify any clear successional gradient (Holler and Abrahamson, 1977; Pitelka, 1977; Turkington and Cavers, 1978; Bostock and Benton, 1979; Harrison, 1979; Primack, 1979; Lovett Doust, 1980a; Waite and Hutchings, 1982). Soule and Werner (1981) point out that the detection of trends in RA along a successional gradient can be dependent on what part of the gradient is studied. We could add that an objective definition of these gradients would also be desirable.

A potential confounding factor in many studies is size-dependent RA. Some of the patterns found can be artifact effects of differences in plant size along the gradient. In some studies, differences in RA between populations could be attributed to the allometric relationship between plant size and RA (Appendix 2).

In some tests of the theory using interspecific (species within the same genus or community-wide) data sets, the support for the predicted decrease in RA with successional age was only due to a change in species composition from annuals in early successional plots to perennials in late successional plots. For example, a re-analysis of the data set of Abrahamson (1979) limiting the contrast to native perennials showed no significant difference in RA between open (10.3 ± 7.0, $n = 14$) and forest (8.2 ± 6.3, $n = 16$) habitats ($t = 0.842$, $df = 28$, $p = 0.588$). Similarly, a re-analysis of the data of Hancock and Pritts (1987) for Asteraceae showed no significant difference in RA across habitats when only perennial species were considered (field: 20.9 ± 11.0, $n = 11$; field and forest: 16.8 ± 10.1, $n = 14$; forest $= 12.7 \pm 12.6$, $n = 5$; Kruskal–Wallis $\chi^2 = 2.362$, $p = 0.307$). In these two cases, however, the trend still was in the predicted direction. Whether or not this "effect of annuals" is considered as a caveat depends on the interpretation of the theory. Although Harper *et al.* (1970) explicitly included the "effect of annuals" in their hypothesis, it is obvious that many of the empirical tests – namely, all the intraspecific ones – have not. Should intraspecific studies be considered as inappropriate tests of the theory?

After screening previous tests in this way, the best agreement with predictions comes from studies of species of the same genus (Fig. 1.3), but even here potential size-dependence of RA, as well as phylogenetic nonindependence could affect the results. As a final remark, all studies presented here have been transversal, i.e., have utilized a space by time substitution. No longitudinal study of successional communities seems to have been attempted.

2. RA and Disturbance Due to the logical connections between disturbance and succession noted earlier, sometimes it is difficult to establish a clear-cut distinction between studies of successional trends and those studying the role of disturbances on RA. As all studies included in the previous section dealt with secondary succession, a disturbance event is implicit in the comparison. However, whereas articles mentioned in the previous section were explicitly framed in a successional context, the ones reviewed in this section will explicitly refer to disturbance regimes within a more or less similar habitat.

Curiously, the classical article by Solbrig and Simpson (1974) showing a differentiation between populations of *Taraxacum officinale* under different disturbance regimes does not present any data on RA. Only absolute number of flower heads are reported, however, this article has been regularly cited as supporting a higher RA in more disturbed environments. Bostock and Benton (1979) found that RA in five perennial Asteraceae agreed with the disturbance regime of their habitat, i.e., species living in more disturbed habitats showed higher RA. Primack (1979) reported higher RA in perennial *Plantago* species growing in disturbed sites as compared as *Plantago* species growing in natural, undisturbed vegetation.

The opposite pattern, i.e., higher RA in less frequently disturbed habitats, has been found by Hartnett (1991) and Reekie (1991). These two studies are remarkable as both included proper replicates of their disturbance regimes, allowed an objective assessment of the gradient studied and, in the case of Hartnett (1991), size-dependence of RA was explicitly accounted for.

A theoretical problem is how to assess the degree of disturbance. As posed by Stearns (1976): "We have few, if any, good data on how stable different environments appear to organisms." Rating of habitats according to disturbance regime is as prone to subjective assessments as successional status. Experimental manipulations of disturbance would improve future experimental designs.

3. Responses of RA to Density The empirical evidence concerning RA and density has been reviewed by Soule and Werner (1981) and Weiner (1988). Density has been the surrogate variable utilized to assess the role of competition on RA. In order not to confound the effect of density with that of other aspects of succession, in this section we will focus on greenhouse studies in which density was manipulated by the researchers, or on studies in which density effects were explicitly separated from other successional aspects (e.g., Abrahamson, 1975b). Once more, we do not intend to be exhaustive in this review, as density has been one of the most studied factors in plant ecology and the literature is vast.

The evidence for a decreased RA in an increasingly competitive (dense) environment is mixed (Fig. 1.4, Appendix 3). However, neglect of size-dependence of RA undermines all these early reports. In fact, the study of size-dependence of RA was promoted after the realization of the considerable

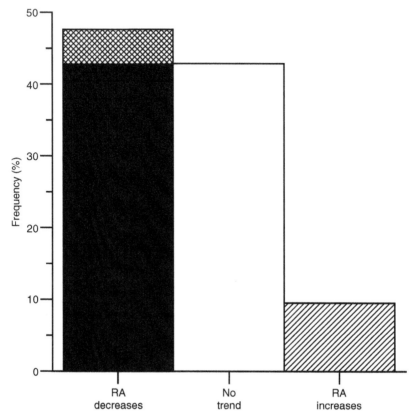

Figure 1.4 Percentage of trials showing decreased (closed bar), similar (open bar), and increased (hatched bar) RA with increasing density ($n = 21$ trials from 14 papers; Appendix 3). The double-hatched bar indicates one study in which flowering was totally suppressed at high density.

plasticity of plant resource allocation in different densities (Samson and Werk, 1986; Ohlson, 1988; Weiner, 1988).

C. RA in Relation to Environmental Stress

Predictions about the response of plants to environmental stress mainly derive from Grime (1979) and his distinction among "competitive" (C), "ruderal" (R), and "stress-tolerant" (S) strategies. According to Grime (1979), "competitive" organisms will show rapid morphogenetic responses to stress which will tend to maximize vegetative growth. In contrast, responses of "stress-tolerant" organisms will be slow and small, and "ruderal" organisms will show a rapid curtailment of vegetative growth and a diversion of resources into flowering. Thus, all the range between decreasing RA, no change and increasing RA in response to stress are expected, depending on the particular strategy (C, S, or R) of a given species.

Testing predictions based on Grime's framework requires that we first fit each species within a strategy. In this kind of "a posteriori" approach (*sensu* Calow and Townsend, 1981), the risk exists of circular reasoning. For example, a decreased RA can be taken as a proof of the plant being a "competitor," when it should be the other way around: one first identifies the plant as a competitor and then tests whether the prediction of a decreased RA is fulfilled. The same caveat applies to r- and K-strategies, in which many times it was difficult to separate predictions from tests of those predictions, due to the empirical (a posteriori) way of description of the strategies. A flavor of circularity can be found in the attempts of Hancock and Pritts (1987) to interpret the data sets of Turkington and Cavers (1978) and Lovett Doust (1980b) within the frame of the r−K theory. C−S−R and r−K theories are not completely tautological, though. One can manage to assign a strategy to a species based on a set of traits and then test if RA patterns fit into the predictions done for that category. However, a more promising approach is taken by Chiariello and Gulmon (1991). These authors have made predictions about the response to stress of RA patterns in annuals and perennials. Those categories, partly related to r−K or C−S−R strategies, are considerably more operative.

Intraspecific responses of RA to stress have been reviewed by Soule and Werner (1981) and, more recently by Chiariello and Gulmon (1991). In the later review, they note that perennial species show a trend to decrease RA in response to resource (light, nutrients, water, length of the growing season) limitation. On the other hand, annuals show an increased RA in response to shortage of some resources (water), but decreased RA in response to others (light), and still no response to others (nutrients) (Chiariello and Gulmon, 1991). As shown in Appendix 4, those predictions are only partially fulfilled and a high variability in responses is shown across factors of stress and life histories.

In the future, attention should be paid to the extent to which responses to stress are adaptive or mere disruptions of the normal resource allocation patterns.

D. Genetic Variation in RA

Soule and Werner (1981) reviewed intraspecific variation for RA and found values within 3−36% of average RA for a given species. To have an evolutionary meaning, that variation should have a genetic basis (Willson, 1983; Bazzaz and Ackerly, 1992). During the 1970s and 1980s, several of the studies mentioned in the previous sections included "common garden" designs aiming at disentangling environmental from genetic influences on the patterns of RA. A partial review of that evidence, in relation to successional maturity, can be found in Hancock and Pritts (1987).

The evidence available shows that in several instances, patterns of RA have a genetic component, i.e., differences in RA between populations are still present when individuals of those populations are grown in a common, controlled environment. However, a substantial amount of phenotypic plasticity has also been found, e.g., associated with changes in plant size. The adaptive or nonfunctional meaning of such plastic responses has been contentious (Weiner, 1988). The existence of such plasticity does not mean that RA cannot be modified by selection. In fact, genetic differences in size-dependent RA have been found (Aarssen and Clauss, 1992; Schmid and Weiner, 1993). As pointed out by Aarssen and Taylor (1992), different patterns of size-dependence in RA can affect plant fitness depending on the competitive environment faced by the individuals. A genetic basis for those patterns means that natural selection could potentially shape them in an adaptive way. This and other aspects of the phenotypic plasticity in RA should be addressed in future studies.

E. What does the Evidence Say?

The review clearly shows several points:

(1) Not all patterns of variation of RA have received the same attention. In particular, the relation between RA and juvenile versus adult mortality deserves further work. On the other hand, a lot of work has been devoted to test patterns of variation of RA with succession, within a theoretical framework (r−K theory) which is now considered outdated. Variation of RA according to stress deserves further theoretical and empirical attention, especially in connection with global change.

(2) Flawed tests of the theory make it difficult to assess whether or not theoretical predictions are supported. Some of these areas would benefit from improvements in the experimental designs and consideration of confounding factors as size-dependent RA and phylogenetic relatedness. Meta-analysis should be applied in future reviews of the evidence, in order to detect lack of power of the tests as well as assess publication bias. Finally, theory should be re-assessed in the light of this mixed empirical support.

VIII. Costs of Reproduction

Costs of reproduction in plants have recently been reviewed by Obeso (2002). The basic assumption underlying most of the research on resource allocation to reproduction is that reproductive investments involve costs and that minimizing such costs or balancing costs, and benefits against each other are important aspects behind the evolution of life histories.

Ultimately the costs are expected in terms of fitness, usually quantified as reduced future fecundity and/or survival. Reproductive costs are, however, also often studied in terms of somatic costs, i.e., reduced growth and/or resource pools of reproductive individuals.

Also, the study of costs of reproduction suffers from a prolific and inconsistently used terminology (Obeso, 2002). Here we will focus mainly on costs in terms of reduced growth or resource pools (i.e., somatic costs) and reduced survival or future fecundity.

A. Methodological Issues

One aspect of reproductive costs that has received much attention is a relatively high frequency of failures to detect costs (cf. Reekie, 1999; Obeso, 2002). Phenotypic correlations in general and observational studies in particular have been suggested as methodological reasons behind failures in detecting costs (e.g., Reznick, 1985, 1992; Partridge and Harvey, 1988; Kozlowski, 1991; Bailey, 1992; Partridge, 1992; Sinervo and Svensson, 1998; Reekie, 1999). Obeso (2002) argued that the proportion of studies failing to detect costs in plants does not vary depending on the methods used. However, comparing the frequencies of failures and successes to detect costs among different types of investigation is associated with several difficulties. For example, the proportion of explained variance tends to be smaller in evolutionary than in ecological or physiological studies (Møller and Jennison, 2002). The risk of committing type II errors (accepting false null hypotheses) may thus be larger in evolutionary than in ecological or physiological studies. Alternatively, the former require larger sample sizes. Furthermore, the tendency to publish studies with accepted null hypotheses probably vary among different types of studies. At this stage we judge it as difficult to compare failure rates in different types of studies.

Although there is an agreement that manipulative studies are strongly preferred before observational, we have seen no attempts to discuss or evaluate potential effects of different means to create nonreproductive individuals. For example, the removal of reproductive parts at an early stage of reproduction may have different effects as compared to using day length or hormone applications to initiate reproduction. Less disruptive manipulations preventing pollination but leaving inflorescences more or less intact measure only a part of the reproductive costs (cf. Jönsson, 2000).

The relatively frequent failures to detect costs have also stimulated many authors to propose biological reasons for why reproductive costs may be smaller than expected or even absent. In the following paragraphs we will shortly summarize those that relate to the allocation of resources (see Obeso, 2002 for a more detailed review; and van Noordwijk and de Jong, 1986; Pease and Bull, 1988; de Laguerie *et al.,* 1991; Houle, 1991; Bailey, 1992 and Stearns, 1992 for other arguments not involving resource allocation).

- Compensatory mechanisms enabling an enhanced resource acquisition during reproduction. The compensatory mechanisms discussed are almost exclusively related to carbon economy. The potential ability to compensate for nutrient investments in reproduction is mainly unknown. The physiological mechanisms behind compensation could be either an increased acquisition in somatic parts (by an increased leaf area or increased carbon dioxide assimilation efficiency) or through carbon assimilation in reproductive parts.
- A temporal separation between reproductive and somatic allocation has been proposed to reduce the reproductive costs (Reekie and Bazzaz, 1992; Ågren and Willson, 1994; Houle, 2001). Similar arguments were used by Dafni *et al.* (1981) when discussing the function of storage in geophytes. This assumes that growth and reproduction are supported by current incomes (strict income breeding, preceeding) and that resources cannot be stored for future use.
- Differential resource needs of somatic and reproductive growth can be another reason for reduced costs (Houle, 2001; Reekie *et al.,* 2002). Also here, the "limiting resource" assumption is then not met if the different functions have different requirements. Since more or less all plant functions need a similar set of resources, only their proportions may vary, different requirements can at the best reduce costs but do not eliminate them.

The extent to which reproduction is resource versus pollen limited has been discussed by many authors (cf. Burd, 1994). Even if reproduction mainly is pollen limited, costs of reproduction caused by resource investments in reproduction can be expected if other aspects of plant performance are resource limited. Similar arguments can be used regarding resource versus meristem limitation of reproduction. If the number of meristems are limiting, trade-offs may occur depending on meristem allocation (see e.g., Obeso, 2002). This does, however, not necessarily exclude resource-based trade-offs. If somatic performance is resource limited, resource mediated trade-offs may also occur.

B. Quantitative Links between Reproductive Allocation and Costs

A depletion of the resource pools of reproductive individuals is commonly regarded as the cause for reproductive costs. Much of the work and discussion on reproductive costs has been qualitative; are there costs or not. We thus largely lack quantitative investigations on how reproductive allocation relates to both somatic and demographic costs. To improve our understanding of resource economy, mediated demographic costs of reproduction, we need quantitative comparisons between reproductive allocation and costs under different situations. For implementing such an approach two theoretical analyses could help resolve some of the controversies discussed above, viz. the concepts of somatic cost of reproduction (*sensu* Tuomi *et al.*, 1983) and that of capital and income breeders.

1. Allocation Systems and the Somatic Cost of Reproduction   As discussed above, the use of energy (biomass) as a currency for measuring reproductive allocation was intially criticized partly due to autotrophic properties of the reproductive organs of many plants. This criticism was, however, based on the assumption that reproductive allocation (then named RE) is a measure, or at least an indicator, of reproductive costs. The fact that reproductive structures often are partly self-supported does not necessarily imply that reproduction or plant performance in general is not limited by energy. The phenomenon of enhanced resource acquisition during reproduction was discussed and analyzed from a theoretical perspective by Tuomi *et al.* (1983). They distinguished between reproductive allocation (they used the term reproductive effort) and reproductive costs and acknowledged that the resource demands of reproduction may be partly or fully met by enhanced resource acquisition. Furthermore, they defined "relative somatic cost" (RSC, Tuomi *et al.*, 1983) as a measure of the cost of reproduction in terms of the resource pool of the reproductive individual:

$$\text{RSC} = \frac{(I_N - I_R)}{I_N}$$

where I_N is resource level in nonreproductive individuals and I_R that of reproductive ones. This somatic cost is later expected to result in reduced survival or future fecundity (Fig. 1.5). If defining reproductive allocation as $RA = I_R / (I_R + I_S)$ (using the symbols defined in Fig. 1.6), RA can be directly compared with the cost (RSC). The original interpretation of RE as a direct measure of a cost corresponds to allocation system A in Fig. 1.6. Here, reproductive allocation occurs exclusively at the expense of somatic allocation and thus RSC = RA. If there are compensatory mechanisms, such as autotrophy in reproductive structures, which fully compensates for reproductive investments, the outcome will be allocation system B1 and there will be no somatic cost of reproduction (RSC = 0). In the extreme case there could even be overcompensation so that the somatic resource pool benefits from reproduction (B2). The final case (B3) of a partial compensation is probably the most common when studying plants and using biomass or energy as currency. Here there is a cost but RSC is smaller than RA. The RSC is then assumed to have direct implications for future survival and/or reproduction. We will come back to the use of these concepts in empirical studies of resource economy.

Tuomi *et al.* (1983) also presented a "threshold hypothesis" where relatively small reproductive efforts (below a threshold) may result in no or even negative survival cost. The argument for this is that "... resource shortage does not automatically mean a shorter life span. On the contrary,

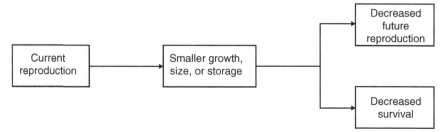

Figure 1.5 Proposed mechanism showing how demographic (future survival and reproduction) costs are mediated by somatic (reduced growth size or storage) costs (modified from Kozlowski, 1991).

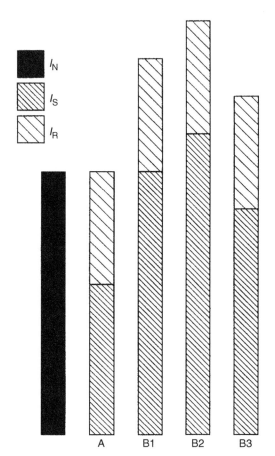

Figure 1.6 Four allocation systems with similar reproductive allocation (I_R) but different somatic allocation (I_S) and compared to somatic investment in a nonreproductive individual (I_N). From Tuomi *et al.* (1983).

moderately starved animals can live even longer than animals feeding *ad libitum*" (Tuomi *et al.*, 1983). In plants, survival is often related to size or growth rate (at least early in life, e.g., Solbrig, 1981b; Thomas and Weiner, 1989; Weih and Karlsson, 1999; but see Wesselingh *et al.*, 1997) and it is uncertain whether this hypothesis is applicable to plants. In any case, in situations when it operates, somatic cost will not result in costs in terms of survival.

Few studies have quantified both RA and RSC, which allow an estimate of the magnitude of the compensatory mechanisms for particular resources. A comparison between RA and RSC for 13 populations (belonging to eight species, Hemborg and Karlsson, 1998) indicates that RSC approximately equals RA for nitrogen and phosphorus (i.e., allocation system A in Fig. 1.6) while RSC tends to be lower than RE in terms of biomass (allocation system B3). For three carnivorous *Pinguicula* species, RSC in terms of both dry matter and N were smaller than RA (calculated from Table 1 in Thorén *et al.*, 1996) while no significant difference was found when using phosphorus allocation. That nitrogen also appears to follow allocation system B3 may be a result of an enhanced resource acquisition through increased prey capture in reproductive individuals (Karlsson *et al.*, 1994). Also Reekie and Bazzaz (1987a,c) estimated RA and RSC for six different genotypes of *Agropyron repens* (Fig. 1.7). In this study, RE_3 corresponds to the "traditional"

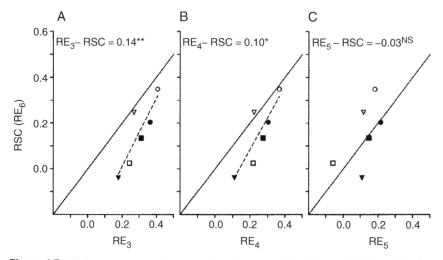

Figure 1.7 Various measures of reproductive allocation (RE_3, RE_4, and RE_5) as related to relative somatic cost of reproduction (RSC or RE_6 in Reekie and Bazzaz, 1987c) for six clones of *Agropyron repens* extracted from figure 4 in Reekie and Bazzaz (1987a) and figure 6 in Reekie and Bazzaz (1987c, results from experiment 1 [open symbols] and 2 [filled symbols], each point is the genotype mean of a series of treatments). RE_3 is the "traditional" reproductive allocation while RE_4 includes also respiration and RE_5 includes respiration and reproductive photosynthesis. The diagonal line indicates a 1:1 relationship between RE and RSC (i.e., allocation system A in Fig. 1.6).

definition of RA (i.e., proportion of reproductive biomass/total biomass) and RE_6 is defined as RSC above. As can be expected from the discussion above, this estimate of RA is significantly higher than the somatic cost of reproduction (RE_6, Fig. 1.7A). Taking respiration in both somatic and reproductive parts into account, the difference between RA (RE_4, Fig. 1.7B) and RSC decreases while it disappears entirely when taking also reproductive photosynthesis into account (RE_5, Fig. 1.7C). This comparison also indicates that the largest differences between RE_3 or RE_4, and RSC are found for genotype-treatment combinations with low RA. Those with a high RA show an RA similar to the somatic cost (allocation system A in Fig. 1.6).

In our view, estimates of RA, RSC, and demographic costs could provide valuable information for our understanding of how reproductive allocation translates into somatic and ultimately demographic costs. Preferentially, the first two (RA and RSC) should be estimated in terms of several different resources.

2. Capital versus Income Breeders   The concepts of income and capital breeding and the partly related concepts of pre- and postbreeding costs have been thoroughly discussed by Jönsson and co-authors (Jönsson, 1997; Jönsson *et al.,* 1995, 1998) and much of this section is based upon these sources. As for many other concepts of reproductive strategies, the concepts of capital and income breeding strategies were developed from studies of birds (Drent and Daan, 1980). Although various authors have defined these terms slightly differently (Jönsson, 1997), a capital breeder is one that uses stored resources for reproduction while an income breeder uses resources captured concurrently with the reproductive activity (Stearns, 1992). These two strategies represent the extreme endpoints of the variation in the degree of temporal separation between acquisition of the resources required for reproduction and the actual reproductive investments, or the extent to which reproductive resource demands are met by storage versus current acquisition.

Storage in plants has been defined as "resources that can be mobilized in the future to support biosynthesis or other plant functions" (Chapin *et al.,* 1990; see also Millard, 1988). With this wide definition, for example, nutrients in the photosynthetic machinery in active leaves can be classified as storage. The translocation of such compounds to reproductive structures, however, would also directly affect somatic functions (leaf photosynthesis). If restricting the term storage to resources not in current active use, such as storage in bulbs, rhizomes, or stems ("reserve storage" *sensu* Chapin *et al.,* 1990), costs in terms of growth or survival may be expected when these reserves are built-up rather than when they are moved from storage sites to reproduction (Jönsson, 1997); "reserve formation directly competes for resources with growth and defence" (Chapin *et al.,* 1990). Empirical support for this is reported for early vegetative stages of *Urtica* (Rosnitschek-Schimmel, 1983, as cited in Lambers *et al.,* 1998). If this is the case, the cost of reproduction may be diluted over a period of time before the apparent onset of reproductive activities while during the season of seed production somatic growth can be maintained without any apparent costs.

The strategy which an individual uses has several important implications in terms of a resource economical analysis of reproduction and the timing of reproductive costs:

(1) There are costs for building up and maintaining storage as well as for the mobilization of the stored reserves and their translocation to sites of use (Penning de Vries, 1975; Chapin *et al.*, 1990). The use of stored resources for reproduction thus involves additional physiological costs for the storage itself, for transformation of storage compounds, and translocation to and from storage sites. These costs are paid during build up and maintenance of storage.

(2) Having stored reserves in stems, roots, or specialized storage organs probably involves an increased value of the storage sites/organs as food items for herbivores (cf. Haukioja *et al.*, 1991). The existence of storage organs may thus increase the risk of losing (parts of) these reserves to herbivores. Alternatively, they may require additional investments for defense.

(3) When resource availability varies over time, such as in seasonal environments, income breeders have to compromise the timing of the reproductive allocation between resource availability, resource demands of other functions, and environmental constraints (e.g., low temperatures) on reproduction. Particularly in cold environments with a short growing season, there may not be much choice of the season when reproductive structures are produced (cf. e.g., Thorén and Karlsson, 1998). Capital breeding relaxes these constraints and allows the successive build up of resources necessary for completing a reproductive event. Furthermore, the use of storage reserves may allow a more rapid production of reproductive (and somatic) structures than may be possible relying on current acquisition alone.

(4) Finally, as discussed above, a capital breeder may face its growth and survival costs when building up storage reserves rather than during reproductive allocation (Fig. 1.8). In that case, survival and growth costs are spread over the same period as the storage build up occurs and no or marginal costs during the actual production of reproductive structures. Alternatively if allocation to storage and reproduction is constant over a reproductive cycle (broken line in Fig. 1.8), the survival cost may be apparently zero. These examples may be speculative, nevertheless they do indicate that costs may not necessarily be expressed when we expect them.

The role of storage for plant reproduction has received little attention (Chapin *et al.*, 1990). Similarly, the concepts of capital and income breeding

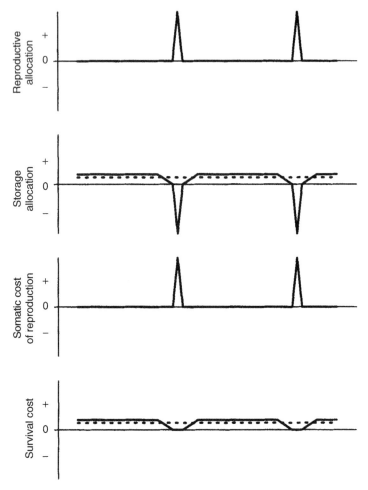

Figure 1.8   Timing of resource allocation to storage and reproduction and cost expression in a capital breeder over two reproductive cycles. Somatic cost (*sensu* Tuomi *et al.,* 1983) is defined in the chapter. Survival cost is assumed to be directly proportional to the resource allocated to storage. The solid line in panel 2 and 4 indicates a hypothetical scenario where allocation to storage is made in between reproductive events while the broken lines indicate a scenario when allocation to storage is constant over time.

have rarely been addressed in botanical studies (Jönsson, 1997). When searching for costs of reproduction these aspects may, however, be of importance. In studies of resource allocation patterns and somatic costs of reproduction (*sensu* Tuomi *et al.,* 1983, cf. above), plants are divided into reproductive and somatic parts. Then, storage reserves are likely included in the somatic part, and then there is no reason to expect large differences in RSC between capital and income breeders. The largest difference is probably some additional costs to the capital breeders for mobilizing and translocating the reserves to and from storage sites. However, when searching for demographic costs in terms of survival, the timing of the cost expression may be entirely or partly uncoupled from the reproductive event (cf. Jönsson *et al.*, 1995; Jönsson, 1997). The fact that prereproductive individuals (i.e., before onset of flowering in plants, classified in a study of reproductive costs as nonreproductive), accumulate resource reserves for reproduction and having costs for this accumulation, may actually have higher costs for building up storage than reproductive individuals have for reproduction. Thus, neglect of capital breeding in an empirical study comparing naturally reproductive and nonreproductive individuals could in the worst case lead to the erroneous conclusion that the cost of reproduction is negative (cf. bottom panel in Fig. 1.8), while actually the prebreeding cost are higher than the cost during actual reproduction. Although the ultimate fate of these resources may be to support reproduction, the proximate cause for a decreased survival in this example is the cost of allocating resources to storage. There is very little empirical evidence available on how capital breeders may react when using experimental means to create nonreproductive individuals. However, for *Lathyrus vernus*, storage and subsequently growth increased after several years of prevented reproduction (Ehrlén and van Groenendael, 2001). Also when using an experimental approach to search for reproductive costs, e.g., by removal of reproductive parts, survival costs may be hidden in a capital breeder's costs if the storage build up occurred before the onset of the experiment.

In terms of carbon, many plants have substantial reserves in roots and other organs (Chapin *et al.*, 1990) that can be a potential source for carbohydrate requirements of reproduction. For other resources recycling of nutrient from senescing leaves or current uptake are probably more important (cf. Chapin *et al.*, 1990). Although some studies have addressed storage in relation to reproduction (e.g., Whigham, 1984; Lubbers and Lechowicz, 1989; Bausenwein *et al.*, 2001), few have addressed storage in relation to cost of reproduction (Ehrlén and van Groenendael, 2001; cf. Jönsson, 1997). There are, however, plants where reproduction and carbon acquisition is entirely separated in time (Mooney and Bartholomew, 1974; Ogden, 1974; Dafni *et al.*, 1981; Pate and Dixon, 1982). How common strict capital breeders are among plants is, however, unknown.

IX. Conclusions

Despite more than 30 years of research in this field and repeated pleas for a resource allocation perspective when studying reproductive strategies and life history phenomena (Harper, 1967; Calow, 1979, 1981, 1994; Solbrig, 1981a; Boggs, 1992, 1994; Jönsson *et al.*, 1995; cf. also Sinervo and

Svensson, 1998), our knowledge of quantitative relationships between resource investments in reproduction and associated costs are largely lacking. If interpreting reproductive allocation (RA) strictly as a descriptive measure, there is no such thing as "wrong currencies." Rather, quantitative comparisons between RA and resource based (i.e., RSC) and/or demographic cost expressions become important for our understanding of the role of resource economics for reproductive costs and thus the evolution of life history strategies. In such contexts, RA and RSC are of interest, irrespective of the resource and ideally several resources should be studied simultaneously. Such comparisons could also be a tool for evaluating cases where no costs are found. As reviewed earlier, there are many suggestions in the literature how plants may reduce or avoid costs caused by a resource depletion of reproductive individuals, but none of these has been thoroughly investigated empirically. Furthermore, it should be kept in mind that constraints other than resource limitations may affect the relationships between RA and costs. For example, exposure to pathogens and herbivory may be affected by reproduction and may thus also affect plant allocation priorities (cf. Michalakis and Hochberg, 1994; Mole, 1994), which in turn may mask resource allocation trade-offs. Thus, from a resource economical perspective on plant reproduction, a major challenge in front of us is to explore quantitative relationships between reproductive allocation, somatic costs, and demographic costs.

Acknowledgments

Johan Erhlén, Ingemar Jönsson, Ed Reekie, Brita Svensson, and Magnus Thorén provided helpful comments on the manuscript.

References

Aarssen L.W. and Clauss M. J. (1992) Genotypic variation in fecundity allocation in *Arabidopsis thaliana. J Ecol* **80**: 109–114.

Aarssen L. W. and Taylor D. R. (1992) Fecundity allocation in herbaceous plants. *Oikos* **65**: 225–232.

Abrahamson W. G. (1975a) Reproduction of *Rubus hispidus* L. in different habitats. *Am Midl Nat* **93**: 471–478.

Abrahamson W. G. (1975b) Reproductive strategies in dewberries. *Ecology* **56**: 721–726.

Abrahamson W. G. (1979) Pattern of resource allocation in wildflower populations of fields and woods. *Amer J Bot* **66**: 71–79.

Abrahamson W. G. and Caswell H. (1982) On the comparative allocation of biomass, energy, and nutrients in plants. *Ecology* **63**: 982–991.

Abrahamson W. G. and Gadgil M. (1973) Growth form and reproductive effort in goldenrods (*Solidago*, Compositae). *Am Nat* **107**: 651–661.

Abrahamson W. G. and Hershey B. J. (1977) Resource allocation and growth of *Impatiens capensis* (Balsaminaceae) in two habitats. *Bull Torrey Bot Club* **104**: 160–164.

Abul-Fatih H. A., Bazzaz F. A. and Hunt R. (1979) The biology of *Ambrosia trifida* L. III. Growth and biomass allocation. *New Phytol* **83**: 829–838.

Ågren G. I. (1985) Limits to plant production. *J theor Biol* **113**: 89–92.

Ågren J. (1988) Sexual differences in biomass and nutrient allocation in the dioecious *Rubus chamaemorus*. *Ecology* **69**: 962–973.

Ågren J., Danell K., Elmqvist T., Ericson L. and Hjältén J. (1999) Sexual dimorphism and biotic interactions. In *Gender and Sexual Dimorphism in Flowering Plants* (Geber M. A., Dawson T. E. and Delph L. F. eds.) pp. 217–246, Springer-Verlag, Berlin.

Ågren J. and Willson M. F. (1994) Cost of seed production in the perennial herbs *Geranium maculatum* and *G sylvaticum* - an experimental field study. *Oikos* **70**: 35–42.

Alpert P., Newell E. A., Chu C., Glyphis J., Gulmon S. L., Hollinger D. Y., Johnson N. D., Mooney H. A. and Puttick G. (1985) Allocation to reproduction in the chaparral shrub, *Diplacus aurantiacus*. *Oecologia* **66**: 309–316.

Antonovics J. (1980) Concepts of resource allocation and partitioning in plants. In *The Allocation of Individual Behaviour* (Staddon J. E. R. ed.) pp. 1–35, Academic Press, New York.

Aronson J., Kigel J. and Shmida A. (1993) Reproductive allocation strategies in desert and Mediterranean populations of annual plants grown with and without water stress. *Oecologia* **93**: 336–342.

Ashman T. L. (1994) A dynamic perspective on the physiological cost of reproduction in plants. *Am Nat* **144**: 300–316.

Ashmun J. W., Brown R. L. and Pitelka L. F. (1985) Biomass allocation in *Aster acuminatus*: variation within and among populations over 5 years. *Can J Bot* **63**: 2035–2043.

Augspurger C. K., Geber M. A. and Evans J. P. (1985) Reproductive output and biomass allocation in *Sesbania emerus* in a tropical swamp. *Amer J Bot* **72**: 1136–1143.

Bailey R. C. (1992) Why we should stop trying to measure the cost of reproduction correctly. *Oikos* **65**: 349–352.

Bausenwein U., Millard P., Thornton B. and Raven J. A. (2001) Seasonal nitrogen storage and remobilization in the forb *Rumex acetosa*. *Funct Ecol* **15**: 370–377.

Bazzaz F. A. (1997) Allocation of resources in plants: state of the science and critical questions. In *Plant Resource Allocation* (Bazzaz F. A. and Grace J. eds.) pp. 1–37, Academic Press, San Diego.

Bazzaz F. A. and Ackerly D. D. (1992) Reproductive allocation and reproductive effort in plants. In *Seeds: The Ecology of Regeneration in Plant Communities* (Fenner M. ed.) pp. 1–26, CAB International, Wallingford.

Bazzaz F. A. and Carlson R. W. (1979) Photosynthetic contribution of flowers and seeds to reproduction of an annual colonizer. *New Phytol* **82**: 223–232.

Bazzaz F. A. and Reekie E. G. (1985) The meaning and measurement of reproductive effort in plants. In *Studies in Plant Demography: A festschrift for John L. Harper* (White J. ed.) pp. 373–387, Academic Press, London.

Bazzaz F. A., Ackerly D. D. and Reekie E. G. (2000) Reproductive allocation in plants. In *Seeds: The Ecology of Regeneration in Plant Communities* (Fenner M. ed.) pp. 1–29, CAB International, Wallingford.

Bazzaz F. A., Carlson R. W. and Harper J. L. (1979) Contribution to reproductive effort by photosynthesis of flowers and fruits. *Nature* **279**: 554–555.

Bazzaz F. A., Chiariello N. R., Coley P. D. and Pitelka L. F. (1987) Allocating resources to reproduction and defence. *BioSci* **37**: 58–67.

Benech Arnold R. L., Fenner M. and Edwards P. J. (1992) Mineral allocation to reproduction in *Sorghum bicolor* and *Sorghum halepense* in relation to parental nutrient supply. *Oecologia* **92**: 138–144.

Bloom A. J., Chapin F. S. III. and Mooney H. A. (1985) Resource limitation in plants – An economic analogy. *Annu Rev Ecol Syst* **16**: 363–392.

Boggs C. L. (1992) Resource allocation: exploring connections between foraging and life history. *Funct Ecol* **6**: 508–518.

Boggs C. L. (1994) The role of resource allocation in understanding reproductive patterns. In *Individuals, Populations and Patterns in Ecology* (Leather S. R., Watt A. D., Mills N. J. and Waltes K. F. A. eds.) pp. 25–33, Intercept, Andover.

Bostock S. J. (1980) Variation in reproductive allocation in *Tussilago farfara. Oikos* **34**: 359–363.

Bostock S. J. and Benton R. A. (1979) The reproductive strategies of 5 perennial Compositae. *J Ecol* **67**: 91–107.

Boutin C. and Morisset P. (1988) Étude de la plasticité phénotypique chez le *Chrysanthemum leucanthemum*. I. Croissance, allocation de la biomasse et reprodution. *Can J Bot* **66**: 2285–2298.

Bradbury I. K. and Hofstra G. (1976) The partitioning of net energy resources in two populations of *Solidago canadensis* along a single developmental cycle in southern Ontario. *Can J Bot* **54**: 2449–2456.

Brock M. A. (1983) Reproductive allocation in annual and perennial species of the submerged aquatic halophyte *Ruppia. J Ecol* **71**: 811–818.

Brouillet L. and Simon J. -P. (1979) Resource allocation and phenology of two species of *Aster* (Asteraceae) and their hybrid. *Can J Bot* **57**: 1792–1799.

Burd M. (1994) Bateman's principle and plant reproduction: the role of pollen limitation in fruit and seed set. *Bot Rev* **60**: 83–90

Calow P. (1979) The cost of reproduction – a physiological approach. *Biol Rev* **54**: 23–40.

Calow P. (1981) Resource utilization and reproduction. In *Physiological Ecology: an Evolutionary Approach to Resource Use* (Townsend C. R. and Calow P. eds.) pp. 245–270, Blackwell, Oxford.

Calow P. (1994) From physiological ecology to population and evolutionary ecology with speculation on the importance of storage processes. In *Individuals, Populations and Patterns in Ecology* (Leather S. R., Watt A. D., Mills N. J. and Walters K. F. A. eds.) pp. 349–358, Intercept, Andover.

Calow P. and Townsend C. R. (1981) Energetics, ecology and evolution. In *Physiological Ecology: an Evolutionary Approach to Resource Use* (Townsend C. R. and Calow P. eds.) pp. 3–19, Blackwell, Oxford.

Carr D. J. and Wardlaw I. F. (1965) The supply of photosynthetic assimilates to the grain from the flag leaf and ear of wheat. *Austr J Bio Sci* **18**: 711–719.

Carpenter A. T. and West N. E. (1988) Reproductive allocation in *Artemisia tridentata* ssp. *vaseyana*: effects of dispersion pattern, nitrogen and water. *Bull Torrey Bot Club* **115**: 161–167.

Cartica R. J. and Quinn J. A. (1982) Resource allocation and fecundity of populations of *Solidago sempervirens* along a coastal dune gradient. *Bull Torrey Bot Club* **109**: 299–305.

Chapin F. S. III (1980) The mineral nutrition of wild plants. *Ann Rev Ecol Syst* **11**: 233–260.

Chapin F. S. III (1989) The cost of tundra plant structures: Evaluation of concepts and currencies. *Am Nat* **133**: 1–19.

Chapin F. S. III, Bloom A. J., Field C. B. and Waring R. H. (1987) Plant responses to multiple environmental factors. *BioSci* **37**: 49–57.

Chapin F. S. III, Schulze E. -D. and Mooney H. A. (1990) The storage and economics of storage in plants. *Annu Rev Ecol Syst* **21**: 423–447.

Charnov E. L. and Schaffer W. M. (1973) Life history consequences of natural selection: Cole's result revisited. *Am Nat* **107**: 791–793.

Cheplick G. P. (1989) Nutrient availability, dimorphic seed production and reproductive allocation in the annual grass *Amphicarpum purshii. Can J Bot* **67**: 2514–2521.

Cheplick G. P. (1995) Plasticity of seed number, mass, and allocation in clones of the perennial grass *Amphibromus scabrivalvis. Int J Plant Sci* **156**: 522–529.

Chiariello N. R. and Gulmon S. L. (1991) Stress effects on plant reproduction. In *Response of Plants to Multiple Stresses* (Mooney H. A., Winner W. E., Pell E. J. and Chu E. eds.) pp. 161–188, Academic Press, San Diego, CA.

Cid-Benevento C. R. and Werner P. A. (1986) Local distributions of old-field and woodland annual plant species: demography, physiological tolerances and allocation of biomass of five species grown in experimental light and soil-moisture gradients. *J Ecol* **74**: 857–880.

Clauss M. J. and Aarssen L. W. (1994) Phenotypic plasticity of size-fecundity relationships in *Arabidopsis thaliana*. *J Ecol* **82**: 447–455.

Cockburn A. (1991) *An Introduction to Evolutionary Ecology*. Blackwell, Oxford.

Cody M. L. (1966) A general theory of clutsch size. *Evolution* **20**: 174–184.

Cole L. C. (1954) The population consequences of life history phenomena. *Q Rev Biol* **29**: 103–137.

Comps B., Thiébaut B., Barrière G. and Letouzey J. (1994) Répartition de la biomasse entre organes végétatifs et reproducteurs chez le hêtre européen (*Fagus sylvatica* L) selon le secteur de la couronne et l'âge des arbres. *Ann Sci For* **51**: 11–26.

Dafni A., Cohen D. and Noy-Meir I. (1981) Life-cycle variation in geophytes. *Ann Missouri Bot Gard* **68**: 652–660.

de Laguerie P., Olivieri I., Atlan A. and Gouyon P. -H. (1991) Analytic and simulation models predicting positive genetic correlations between traits linked by trade-offs. *Evol Ecol* **5**: 361–369.

de Ridder F. and Dhondt A. A. (1992a) The demography of a clonal herbaceous perennial plant, the longleaved sundew *Drosera intermedia*, in different heathland habitats. *Ecography* **15**: 129–143.

de Ridder F. and Dhondt A. A. (1992b) The reproductive behaviour of a clonal herbaceous plant, the longleaved sundew *Drosera intermedia*, in different heathland habitats. *Ecography* **15**: 144–153.

DeFalco L. A., Bryla D. R., Smith-Longozo V. and Nowak R. S. (2003) Are Mojave Desert annual species equal? Resource acquisition and allocation for the invasive grass *Bromus matritensis* subsp. *rubens* (Poaceae) and two native species. *Amer J Bot* **90**: 1045–1053.

Douglas D. A. (1981) The balance between vegetative and sexual reproduction of *Mimulus primuloides* (Schrophulariaceae) at different altitudes in California. *J Ecol* **69**: 295–310.

Drent R. H. and Daan S. (1980) The prudent parent: energetic adjustments in avian breeding. *Ardea* **68**: 225–252.

Dunn C. P. and Sharitz R. R. (1991) Population structure, biomass allocation, and phenotypic plasticity in *Murdannia keisak* (Commelinaceae) *Amer J Bot* **78**: 1712–1723.

Eckstein R. L. and Karlsson P. S. (2001) The effect of reproduction on nitrogen use-efficiency of three species of the carnivorous genus *Pinguicula*. *J Ecol* **89**: 798–806.

Ehrlén J. and van Groenendael J. (2001) Storage and the delayed costs of reproduction in the understorey prennial *Lathyrus vernus*. *J Ecol* **89**: 237–246.

Evans L. T. and Rawson H. M. 1970. Photosynthesis and respiration by the flag leaf and components of the ear during grain development in wheat. *Austr J Biol Sci* **23**: 235–254.

Evenson W. E. (1983) Experimental studies of reproductive energy allocation in plants. In *Handbook of Experimental Pollination Biology* (Jones C. E. and Little R. J. eds.) pp. 249–274, Scientific and Academic Editions, New York.

Fenner M. (1985) *Seed Ecology*. Chapman Hall, London.

Fenner M. (1986) The allocation of minerals to seeds in *Senecio vulgaris* plants subjected to nutrient shortage. *J Ecol* **74**: 385–392.

Fisher R. A. (1930) *The Genetical Theory of Natural Selection*. Oxford University Press, Oxford.

Ford E. D. (2000) *Scientific Method for Ecological Research*. Cambridge University Press, Cambridge.

Funk R. W. (1979) *Ecological Energetics of Arctic and Alpine Populations of Saxifraga Cernua L.* PhD thesis, Duke University.

Gadgil M. and Solbrig O. T. (1972) The concept of r- and K-selection: evidence for wild flowers and some theoretical considerations. *Am Nat* **106**: 14–31.

Galen C., Dawson T. E. and Stanton M. L. (1993) Carpels as leaves – meeting the carbon cost of reproduction in an alpine buttercup. *Oecologia* **95**: 187–193.

Gaines M. S., Vogt K. J., Hamrik J. L. and Caldwell J. (1974) Reproductive strategies and growth patterns in sunflowers (*Helianthus*) *Am Nat* **108**: 889–894.

Gleeson S. K. and Tilman D. (1992) Plant allocation and the multiple limitation hypothesis. *Am Nat* **139**: 1322–1343.

Goldman D. A. and Willson M. F. (1986) Sex allocation in functionally hermaphrodictic plants: a review and critique. *Bot Rev* **52**: 157–194.

Grace J. B. and Wetzel R. G. (1981) Effects of size and growth rate on vegetative reproduction in Typha. *Oecologia* **50**: 158–161.

Grime J. P. (1979) *Plant Strategies and Vegetation Processes*. Wiley and Sons, Chichester.

Gross K. L., Berner T., Marschall E. and Tomcko C. (1983) Patterns of resource allocation among five herbaceous perennials. *Bull Torrey Bot Club* **110**: 345–352.

Grulke N. E. and Bliss L. C. (1988) Comparative life history characteristics of two high arctic grasses, Northwest Territories. *Ecology* **69**: 484–496.

Hancock J. F. and Pritts M. P. (1987) Does reproductive effort vary across different life forms and seral environments? A review of the literature. *Bull Torrey Bot Club* **114**: 53–59.

Hansson M. L. (1996) Biomass partitioning and its effect on reproduction in a monocarpic perennial (*Anthriscus sylvestris*) Response to nitrogen and light supply. *Acta Bot Neerl* **45**: 345–354.

Harrison P. G. (1979) Reproductive strategies in intertidal populations of two co-ocurring seagrasses (*Zostera* spp.) *Can J Bot* **57**: 2635–2638.

Harper J. L. (1967) A Darwinian approach to plant ecology. *J Ecol* 347–270.

Harper J. L. (1977) *Population Biology of Plants*. Academic Press, London.

Harper J. L. and Ogden J. (1970) The reproductive strategy of higher plants I. The concept of strategy with special reference to *Senecio vulgaris* L. *J Ecol* **58**: 681–698.

Harper J. L., Lovell P. H. and More K. G. (1970) The shapes and sizes of seeds. *Annu Rev Ecol Syst* **1**: 327–356.

Hartnett D. C. (1991) Effects of fire in tallgrass prairie on growth and reproduction of prairie coneflower (*Rabitida columnifera*: Asteraceae) *Amer J Bot* **78**: 429–435.

Harvey P. H. and Pagel M. (1991) *The Comparative Method in Evolutionary Biology*. Oxford University Press, Oxford.

Haukioja E., Ruohomaki K., Suomela J. and Vuorisalo T. (1991) Nutritional quality as a defence against herbivores. *Forest Ecol Manag* **39**: 237–245.

Hawthorn W. R. and Cavers P. B. (1978) Resource allocation in young plants of two perennial species of *Plantago*. *Can J Bot* **56**: 2533–2537.

Hemborg Å. M. and Karlsson P. S. (1998) Somatic costs of reproduction in eight subarctic plant species. *Oikos* **82**: 149–157.

Hemborg Å. M. and Karlsson P. S. (1999) Sexual differences in biomass and nutrient allocation of first-year *Silene dioica* plants. *Oecologia* **118**: 453–460.

Hermanutz L. A. and Weaver S. E. (1996) Agroecotypes or phenotypic plasticity? Comparison of agrestal and ruderal populations of the weed *Solanum ptycanthum*. *Oecologia* **105**: 271–280.

Hickman J. C. (1975) Environmental unpredictability and plastic energy allocation strategies in the annual *Polygonum cascadense* (Polygonaceae). *J Ecol* **63**: 689–701.

Hickman J. C. (1977) Energy allocation and niche differentiation in four co-existing annual species of *Polygonum* in western North America. *J Ecol* **65**: 317–326.

Hickman J. C. and Pitelka L. F. (1975) Dry weight indicates energy allocation in ecological strategy analysis of plants. *Oecologia* **21**: 117–121.

Holler J. R. and Abrahamson W. G. (1977) Seed and vegetative reproduction in relation to density in *Fragaria virginiana* (Rosaceae). *Amer J Bot* **64**: 1003–1007.

Houle D. (1991) Genetic covariance of fitness correlates: what genetic correlations are made of and why it matters. *Evolution* **45**: 630–648.

Houle G. (2001) Reproductive costs are associated with both the male and female functions in *Alinus viridis* ssp. *crispa Écoscience* **8**: 220–229.

Jackson R. G. and Caldwell M. M. (1993) The scale of nutrient heterogeneity around individual plants and its quantification with geostatistics. *Ecology* **74**: 612–614.

Jaksic F. M. and Montenegro G. (1979) Resource allocation of Chilean herbs in response to climatic and microclimatic factors. *Oecologia* **40**: 81–89.

Jolls C. L. (1984) Contrasting resource allocation patterns in *Sedum lanceolatum* Torr: Biomass versus energy estimates. *Oecologia* **63**: 57–62.

Jönsson K. I. (1997) Capital and income breeders as alternative tactics of resource use in reproduction. *Oikos* **68**: 57–66.

Jönsson K. I. (2000) Life history consequences of fixed costs of reproduction. *Écoscience* **7**: 423–427.

Jönsson K. I. and Tuomi J. (1994) Costs of reproduction in a historical perspective. *Trends Ecol Evol* **9**: 304–307.

Jönsson K. I., Tuomi J. and Järemo J. (1995) Reproductive effort tactics: balancing pre- and postbreeding costs of reproduction. *Oikos* **74**: 35–44.

Jönsson K. I., Tuomi J. and Järemo J. (1998) Pre-and postpreding costs of parental investment. *Oikos* **83**: 424–431.

Jurik T. W. (1983) Reproductive effort and CO_2 dynamics of wild strawberry populations. *Ecology* **64**: 1329–1342.

Jurik T. W. (1985) Differential costs of sexual and vegetative reproduction in wild strawberry populations. *Oecologia* **66**: 394–403.

Karlsson P. S. (1994) Photosynthetic capacity and photosynthetic nutrient use efficiency of *Rhododendron lapponicum* leaves as related to leaf nutrient status, leaf age and branch reproductive status. *Funct Ecol* **8**: 694–700.

Karlsson P. S. and Pate J. S. (1992) Resource allocation to asexual gemma production and sexual reproduction in south west Australian pygmy and micro stilt-form species of sundew (*Drosera* spp, Droseraceae) *Austr J Bot* **40**: 353–364.

Karlsson P. S., Svensson B. M., Carlsson B. Å. and Nordell K. O. (1990) Resource investments in reproduction and their consequences for three *Pinguicula* species. *Oikos* **59**: 393–398.

Karlsson P. S., Thorén M. and Hanslin H. M. (1994) Patterns of prey capture by three *Pinguicula* species in a subarctic environment. *Oecologia* **99**: 188–193.

Kawano S. and Masuda J. (1980) The productive and reproductive biology of flowering plants. VII. Resource allocation and reproductive capacity in wild populations of *Heloniopsis orientalis* (Thunb.) C. Tanaka (Liliaceae). *Oecologia* **45**: 307–317.

Keeley J. E. and Keely S. C. (1977) Energy allocation patterns of a sprouting and a nonsprouting *Arctostaphylos* species in the California chaparral. *Am Midl Nat* **98**: 1–10.

Kozlowski J. (1991) Optimal energy allocation models – an alternative to the concepts of reproductive effort and cost of reproduction. *Acta Oecol* **12**: 11–33.

Kobayashi T., Ikeda H. and Hori Y. (1999) Growth analysis and reproductive allocation of Japanese forbs and grasses in relation to organ toughness under trampling. *Pl Biol.* **1**: 445–452.

Kozlowski T. T. (1971) *Growth and Development of Trees. Vol II. Cambial Growth, Root Growth, and Reproductive Growth.* Academic Press, New York.

Lack D. (1954) *The Natural Regulation of Animal Numbers.* Claredon Press, Oxford.

Lambers H., Chapin F. S. III. and Pons T. L. (1998) *Plant Physiological Ecology.* Springer-Verlag, New York.

Laporte M. M. and Delph L. F. (1996) Sex-specific physiology and source-sink relations in the dioecious plant *Silene latifolia*. *Oecologia* **106**: 63–72.

Leonard E. R. (1962) Inter-relations of vegetative and reproductive growth, with special reference to indeterminate plants. *Bot Rev* **28**: 353–410.

Li B., Shibuya T., Yogo Y., Hara T. and Matsuo K. (2001a) Effects of light quantity and quality on growth and reproduction of a clonal sedge, *Cyperus sculentus*. *Pl Sp Biol* **16**: 69–81.

Li B., Shibuya T., Yogo Y., Hara T. and Yokozawa M. (2001b) Interclonal differences, plasticity and trade-offs of life history traits of *Cyperus sculentus* in relation to water availability. *Pl Sp Biol* **16**: 193–207.

Liebig J. (1855) *Die Grundsatze der Agrochemie*. Viewig, Brauns.

Lovett Doust J. (1980a) A comparative study of the life history and resource allocation in selected Umbeliferae. *Biol J Linn Soc* **13**: 139–154.

Lovett Doust J. (1980b) Experimental manipulation of patterns of resource allocation in the growth cycle and reproduction of *Smyrnium olusatrum* L. *Biol J Linn Soc* **13**: 155–166.

Lovett Doust J. (1989) Plant reproductive strategies and resource allocation. *Trends Ecol Evol* **4**: 230–234.

Lubbers A. E. and Lechowicz M. J. (1989) Effects of leaf removal on reproduction vs. belowground storage in *Trillium grandiflorum*. *Ecology* **70**: 86–96.

Luftensteiner H. W. (1980) Der Reproduktionsaufwand in vier mitteleuropäischen Plfanzengemeinschaften. *Pl Syst Evol* **135**: 235–251.

Marshall C. and Watson M. A. (1992) Ecological and physiological aspects of reproductive allocation. In *Fruit and Seed Production* (Marshall C. and Grace J. eds.) pp. 173–202, Cambridge University Press, Cambridge.

Marschner H. (1995) *Mineral Nutrition of Higher Plants*. Academic Press, London.

Mattirolo O. (1899) Sulla influenza che la estirpazione die fiori esercita sui tubercoli radicali delle piante leguminose. *Malphighia* **13**: 382–421.

Maun M. A. (1974) Reproductive biology of *Rumex crispus*. Phenology, surface area of chlorophyll containing tissue and contribution of perianth to reproduction. *Can J Bot* **52**: 2181–2187.

McDowell S. C. L., McDowell N. G., Marshall J. D. and Hultine K. (2000) Carbon and nitrogen allocation to male and female reproduction in Rocky Mountain Douglas-fir (*Pseudotsuga menziesii* var. *glauca*, Pinaceae) *Amer J Bot* **87**: 539–546.

McNamara J. and Quinn J. A. (1977) Resource allocation and reproduction in populations of *Amphicarpum purshii* (Gramineae) *Amer J Bot* **64**: 17–23.

Michalakis Y. and Hochberg M. E. (1994) Parasitic effects on host life-history traits: a review of recent studies. *Parasite* **1**: 291–294.

Millard P. (1988) The accumulation and storage of nitrogen by herbaceous plants. *Plant Cell Environm* **11**: 1–8.

Mole S. (1994) Trade-offs and constraints in plant-herbivore defence theory: a life-history perspective. *Oikos* **71**: 3–12.

Møller A. P. and Jennions M. D. (2002) How much variance can be explained by ecologists and evolutionary biologists. *Oecologia* **132**: 492–500.

Mooney H. A. and Bartholomew B. (1974) Comparative carbon balance and reproductive moodes of two Californian *Aesculus* species. *Botanical Gazette* **135**: 303–313.

Newell S. J. and Tramer E. J. (1978) Reproductive strategies in herbaceous plant communities during succession. *Ecology* **59**: 228–234.

Obeso J. R. (2002) The cost of reproduction in plants. *New Phytol* **155**: 321–348.

Obeso J. R., Álvarez-Santullano M. Retuerto R. (1998) Sex ratios, size distributions, and sexual dimorphism in the dioecious tree *Ilex aquifolium* (Aqufoliaceae). *Amer J Bot* **85**: 1602–1608.

Odum E. P. (1961) *Fundamentals of Ecology*. W. B. Saunders Company, Philadelphia and London.

Ogden J. (1974) The reproductive strategy of higher plants II. The reproductive strategy of *Tussilago farara* L. *J Ecol* **62**: 291–324.

Ohlson M. (1988) Size-dependent reproductive effort in three populations of *Saxifraga hirculus* in Sweden. *J Ecol* **76**: 1007–1016.

Ong J. K., Colville K. E. and Marshall C. (1978) Assimilation of $^{14}CO_2$ by the inflorescence of *Poa annua* and *Lolium perenne* L. *Ann Bot* **42**: 855–862.

Pate J. S. and Dixon K. W. (1982) *Tuberous, Cormous and Bulbous Plants. Biology of an Adaptive Strategy in Western Australia.* University of Western Australia Press, Perth.

Partridge L. (1992) Measuring reproductive costs. *Trends Ecol Evol* **7**: 99–100.

Partridge L. and Harvey P. H. (1985) Costs of reproduction. *Nature* **316**: 20.

Pease C. M. and Bull J. J. (1988) A critique of methods for measuring life history trade-offs. *J Evol Biol* **1**: 293–303.

Penning de Vries F. W. T. (1975) The cost of maintenance processes in plant cells. *Ann Bot* **39**: 77–92.

Piñero D., Sarukhán J. and Alberdi P. (1982) The cost of reproduction in a tropical palm, *Astrocaryum mexicanum. J Ecol* **70**: 473–481.

Pitelka L. F. (1977) Energy allocation in annual and perennial lupines (*Lupinus*: Leguminosae) *Ecology* **58**: 1055–1065.

Pitelka L. F., Stanton D. S. and Peckenham M. O. (1980) Effects of light and density on resource allocation in a forest herb, *Aster acuminatus* (Compositae) *Amer J Bot* **67**: 942–948.

Pratt C. R. Jr (1984) The response of *Solidago graminifolia* and *S. juncea* to nitrogen fertilizer application: changes in biomass allocation and implications for community structure. *Bull Torrey Bot Club* **111**: 469–478.

Primack R. B. (1979) Reproductive effort in annual and perennial species of *Plantago* (Plantaginaceae) *Am Nat* **114**: 51–62.

Primack R. B. and Antonovics J. (1982) Experimental ecological genetics in *Plantago*. VII. Reproductive effort in populations of *Plantago lanceolata* L. *Evolution* **36**: 742–752.

Primack R. B., Rittenhouse A. R. and August P. V. (1981) Components of reproductive effort and yield in goldenrods. *Amer J Bot* **68**: 855–858.

Pritts M. P. and Hancock J. F. (1983) Seasonal and lifetime allocation patterns of the woody goldenrod, *Solidago pauciflosculosa* Micaux. (Compositae) *Amer J Bot* **70**: 216–221.

Pritts M. P. and Hancock J. F. (1985) Lifetime biomass partitioning and yield component relationships in the highbush blueberry, *Vaccinium corymbosum* L. (Ericaceae) *Amer J Bot* **72**: 446–452.

Reekie E. G. (1991) Cost of seed versus rhizome production in *Agropyron repens. Can J Bot* **69**: 2678–2683.

Reekie E. G. (1997) Trade-offs between reproduction and growth influence time of reproduction. In *Plant Resource Allocation.* (Bazzaz F. A. and Grace J. eds.) pp. 191–209, Academic Press, San Diego, CA

Reekie E. G. (1999) Resource allocation, trade-offs, and reproductive effort in plants. In *Life History Evolution in Plants.* (Vourisalo T. O. and Mutikainen P. K. eds.) pp. 173–193, Kluwer Academic Publishers, Dordrecht.

Reekie E. G. and Bazzaz F. A. (1987a) Reproductive effort in plants. 1. Carbon allocation to reproduction. *Am Nat* **129**: 867–896.

Reekie E. G. and Bazzaz F. A. (1987b) Reproductive effort in plants. 2. Does carbon reflect the allocation of other resources? *Am Nat* **129**: 897–906.

Reekie E. G. and Bazzaz F. A. (1987c) Reproductive effort in plants. 3. Effects of reproduction on vegetative activity. *Am Nat* **129**: 907–919.

Reekie E. G. and Bazzaz F. A. (1992) Cost of reproduction as reduced growth in genotypes of two congeneric species with constrasting life histories. *Oecologia* **90**: 21–26.

Reekie E. G., Budge S. and Baltzer J. L. (2002) The shape of the trade-off function between reproduction and future performance in *Plantago major* and *Plantago rugelii. Can J Bot* **80**: 140–150.

Reznick D. (1985) Cost of reproduction: an evaluation of the empirical evidence. *Oikos* **44**: 257–267.

Reznick D. (1992) Measuring the costs of reproduction. *Trends Ecol Evol* **7**: 42–45.

Reznick D., Briant M. J. and Bashey F. (2002) r and K-selection revisisted: the role of population regulation in life-history evolution. *Ecology* **83**: 1509–1520.

Roff D. A. (1992) *The Evolution of Life Histories: Theory and Analysis*. Chapman and Hall, New York.

Rosnitschek-Schimmel I. (1983) Biomass and nitrogen partitioning in a perennial and an annual nitrophilic species of *Urtica. Z Pflanzenphysiol* **109**: 215–225.

Roos F. H. and Quinn J. A. (1977) Phenology and reproductive allocation in *Andropogon scoparius* (Graamineae) populations in communities of different successional stages. *Amer J Bot* **64**: 535–540.

Rubio G., Zhu J. and Lynch J. P. (2003) A critical test of the two prevailing theories of plant responses to nutrient availability. *Amer J Bot* **90**: 143–152.

Salisbury E. J. 1942. *The Reproductive Capacity of Plants*. G. Bell and Sons, London.

Samson D. A. and Werk K. S. (1986) Size-dependent effects in the analysis of reproductive effort in plants. *Am Nat* **127**: 667–680.

Sarukhán J. (1974) Studies on plant demography: *Ranunculus repens* L., *R. bulbosus* L. and *R acris* L. II. Reproductive strategies and seed population dynamics. *J Ecol* **62**: 151–177.

Schaffer W. M. (1974) Optimal reproductive effort in fluctuating environments. *Am Nat* **108**: 783–790.

Schaffer W. M. and Gadgil M. P. (1977) Selection for optimal life histories in plants. In *Ecology and Evolution of Communities*. 2nd ed. (Cody M. L. and Diamond J. M. eds.) pp. 142–157, Belkamp Press of Harvard University Press, Cambridge, MS.

Schat H. Ouborg J. and de Wit R. (1989) Life history and plant architecture: size-dependent reproductive allocation in annual and biennial *Centaurium* species. *Acta Bot Neerl* **38**: 183–201.

Schmid B. and Weiner J. (1993) Plastic relationships between reproductive and vegetative mass in *Solidago altissima. Evolution* **47**: 61–74.

Silvertown J. and Dodd M. (1997) Comparing plants and connecting traits. In *Plant Life Histories: Ecology, Phylogeny and Evolution*. (Silvertown J., Franco M. and Harper J. L. eds.) pp. 3–16, Cambridge University Press, Cambridge.

Sinervo B. and Svensson E. (1998) Mechanistic and selective causes of life history trade-offs and plasticity. *Oikos* **83**: 432–442.

Snell T. W. and Burch D. G. (1975) The effects of density on resource partitioning in *Chamaesyce hirta. Ecology* **56**: 742–746.

Sobrado M. A. and Turner N. C. (1986) Photosynthesis, dry matter accumulation and distribution in the wild sunflower *Helianthus petiolaris* and the cultivated sunflower *Helianthus annuus* as influenced by water deficits. *Oecologia* **69**: 181–187.

Solbrig O. T. (1971) The population biology of dandelions. *Am Sci* **59**: 686–694.

Solbrig O. T. (1980). Demography and natural selection. In *Demography and Evolution in Plant Populations*. (Solbrig O. T. ed.) pp. 1–20, Blackwell Scientific Publications, Oxford.

Solbrig O. T. (1981a) Energy, information and plant evolution. In *Physiological Ecology: An Evolutionary Approach to Resource Use*. (Townsend C. R. and Calow P. eds.) pp. 274–299, Blackwell, Oxford.

Solbrig O. T. (1981b) Studies on the population biology of the genus Viola. II. The effect of plant size on fitness in *Viola sororia. Evolution* **35**: 1080–1093.

Solbrig O. T. and Simpson B. B. (1974) Components of regulation of a population of dandelions in Michigan. *J Ecol* **62**: 473–486.

Soule J. D. and Werner P. A. (1981) Patterns of resource allocation in plants, with special reference to *Potentilla recta* L. *Bull Torrey Bot Club* **108**: 311–319.

Southwick E. E. (1984) Photosynthetic allocation to floral nectar: A neglected investment. *Ecology* **65**: 1775–1779.

Spira T. P. and Pollack O. D. (1986) Comparative reproductive biology of alpine biennial and perennial gentians (*Gentiana*: Gentianaceae) in California. *Amer J Bot* **73**: 39–47.

Stearns S. C. (1976) Life-history tactics: A review of the ideas. *Q Rev Biol* **51**: 3–47.

Stearns S. C. (1992) *The Evolution of Life Histories*. Oxford University Press, Oxford.

Stewart A. J. A. and Thompson K. (1982) Reproductive strategies of six herbaceous perennial species in relation to a successional sequence. *Oecologia* **52**: 269–272.

Struik G. O. (1965) Growth patterns in some native annual and perennial herbs in southern Wisconsin. *Ecology* **46**: 401–420.

Svensson B. M., Carlsson B. Å., Karlsson P. S. and Nordell K. O. (1993) Population dynamics of three *Pinguicula* species in a subarctic environment. *J Ecol* **81**: 635–645.

Thomas S. C. and Weiner J. (1989) Growth, death and size distribution change in an *Impatiens pallida* population. *J Ecol* **77**: 524–536.

Thompson K. and Stewart A. J. A. (1981) The measurement and meaning of reproductive effort in plants. *Am Nat* **117**: 205–211.

Thomson D. L. (1996) Age-specific life history tactics in organisms with indeterminate growth: optimal models for non-optimal behavior? *Ecol Res* **11**: 61–68.

Thorén L. M. and Karlsson P. S. (1998) Effects of supplementary feeding on growth and reproduction of three carnivorous plant species in a subarctic environment. *J Ecol* **86**: 501–510.

Thorén L. M., Karlsson P. S. and Tuomi J. (1996) Relationships between reproductive effort and relative somatic cost of reproduction for three *Pinguicula species*. *Oikos* **76**: 427–434.

Thorne G. N. (1963) Varietal differences in photosynthesis of ears and leaves of barley. *Ann Bot* **27**: 155–174.

Thorne G. N. (1965) Photosynthesis of ears and flag leaves on wheat and barley. *Ann Bot* **29**: 317–329.

Tuomi J., Hakala T. and Haukioja E. (1983) Alternative concepts of reproductive effort, cost of reproduction and selection in life-history evolution. *Amer Zool* **23**: 25–34.

Turkington R. A. and Cavers P. B. (1978) Reproductive strategies and growth patterns in four legumes. *Can J Bot* **56**: 413–416.

van Andel J. and Vera F. (1977) Reproductive allocation in *Senecio sylvaticus* and *Chamenerion angustifolium* in relation to mineral nutrition. *J Ecol* **65**: 747–758.

van Noordwijk A. J. and de Jong J. (1986) Acquisition and allocation of resources: their influence on variation in life history tactics. *Am Nat* **128**: 137–142.

Verburg R. and Grava D. (1998) Differences in allocation patterns in clonal and sexual offspring in a woodland pseudo-annual. *Oecologia* **115**: 472–477.

Waite S. and Hutchings M. J. (1982) Plastic allocation patterns in *Plantago coronopus*. *Oikos* **38**: 333–342.

Watson D. J., Thorne G. N. and French S. A. W. (1963) Analysis of growth and yield of winter and spring wheats. *Ann Bot* **27**: 1–22.

Watson M. A. and Casper B. B. (1984) Morphogenetic constraints on patterns of carbon distribution in plants. *Annu Rev Ecol Syst* **15**: 233–258.

Weaver S. E. and Cavers P. B. (1980) Reproductive effort of two perennial weed species in different habitats *J Appl Ecol* **17**: 505–513.

Weih M. and Karlsson P. S. (1999) The nitrogen economy of mountain birch seedlings: implications for winter survival. *J Ecol* **87**: 211–219.

Weiner J. (1988) The influence of competition on plant reproduction. In *Plant Reproductive Ecology: Patterns and Strategies*. (Lovett Doust J. and Lovett Doust L. eds.) pp. 228–245, Oxford University Press, Oxford.

Weiss P. V. (1978) Reproductive efficiency and growth of *Emex australis* in relation to stress. *Aust J Ecol* **3**: 57–65.

Wesselingh R. A., Klinkhamer P. G. L., de Jong T. J. and Boorman L. A. (1997) Threshold size for flowering in different habitats: effects of size-dependent growth and survival. *Ecology* **78**: 2118–2132.

Whigham D. F. (1984) Biomass and nutrient allocation of *Tipularia discolor* (Orchidaceae) *Oikos* **42**: 303–313.

Williams G. C. (1966a) Natural selection, the cost of reproduction, and a refinement of Lacks's principle. *Am Nat* **100**: 687–690.

Williams G. C. (1966b) *Adaptation and Natural Selection: A Critique of some Current Evolutionary Thought.* Princeton University Press, Princeton, New Jersey.

Williams K., Koch G. W. and Mooney H. A. (1985) The carbon balance of flowers of *Diplacus aurantiacus. Oecologia* **66**: 530–535.

Willson M. F. (1983) *Plant Reproductive Ecology.* Wiley and Sons, New York.

Wilson A. M. and Thompson K. (1989) A comparative study of reproductive allocation in 40 British grasses. *Funct Ecol* **3**: 297–302.

Winn A. A. and Pitelka L. F. (1981) Some effects of density on the reproductive patterns and patch dynamics of *Aster acuminatus. Bull Torrey Bot Club* **108**: 438–445.

Young D. R. (1983) Comparison of intraspecific variations in the reproduction and photosynthesis of an understory herb, *Arnica cordifolia. Amer J Bot* **70**: 728–734.

Appendix 1

Main Data Sets Utilized to Test the Prediction of a Higher RE in Semelparous Species Compared to Iteroparous Ones

Reference	Scope	Support	Size-dependence?
Struik (1965)	Whole community	Yes	No
Gaines *et al.* (1974)	4 *Helianthus* spp.	Yes	Yes
Pitelka (1977)	3 *Lupinus* spp.	Yes	No
Turkington and Cavers (1978)	2 *Medicago* spp.	Yes	No
Abrahamson (1979)	Whole community	Yes?[a]	No
Primack (1979)	15 *Plantago* spp.	Yes	Yes?
Luftensteiner (1980)	Whole community	Yes	No
Brock (1983)	3 *Ruppia* spp.	Yes	No
Spira and Pollack (1986)	3 *Gentiana* spp.	Yes	No
Benech Arnold *et al.* (1992)	2 *Shorgum* spp.	Yes	No

[a] Trend in the right direction, but results only marginally significant.

Appendix 2

Main Data Sets Testing the Prediction of a Higher RA in Early Successional Habitats Compared to Late Successional Ones. References have mostly been taken from Hancock and Pritts (1987), with a few Additions, Corrections, and exclusions, as Justified in the Text

Reference	Taxa – life history[a]	Populations[b]	Succession data[c]	Support?[d]	Size included? (assessment)[e]
Single Species					
Abrahamson and Gadgil (1973)	*Solidago rugosa* – P	2	Ph (open to forest)	Yes	Yes (influential)
Abrahamson and Gadgil (1973)	*Solidago speciosa* – P	2	Ph (open to forest)	Yes	Yes (influential)
Abrahamson (1975a)	*Rubus hispidus* – P	3	Ph (open to forest)	Yes	No (differences in size)
Abrahamson (1975b)	*Rubus trivialis* – P	3	Ph (all with tress)	No	No (differences in size)
Hickman (1975)	*Polygonum cascadense* – A	5	Ph + Env (all open?)	Yes	No (negative sd RA?)
Bradbury and Hofstra (1976)	*Solidago canadensis* – P	2	Ph (pastures)	Right trend	No (positive sd RA?)
Hickman (1977)	*Polygonum douglasii* –A	3	Ph + Env (all open?)	No	No (si RA)
Hickman (1977)	*Polygonum kelloggii* – A	3	Ph + Env (all open?)	No	No (si RA)
Hickman (1977)	*Polygonum minimum* – A	2	Ph + Env (all open?)	No	No (si RA)
Keeley and Keeley (1977)	*Arctostaphyllos glandulifera* – P	2	Age (chaparral)	No	No
Keeley and Keeley (1977)	*Arctostaphyllos glauca* – P	2	Age (chaparral)	No	No
McNamara and Quinn (1977)	*Amphicarpum purshii* – A	5	Age (ditches)	No	No
Roos and Quinn (1977)	*Andropogon scoparius* – P	6 (1 rep)	Age (open to bushy)	? (1 odd)	Similar size was chosen

(Continued)

Appendix 2

Main Data Sets Testing the Prediction of a Higher RA in Early Successional Habitats Compared to Late Successional Ones. References have mostly been taken from Hancock and Pritts (1987), with a few Additions, Corrections and exclusions, as Justified in the Text—cont'd

Reference	Taxa – life history[a]	Populations[b]	Succession data[c]	Support?[d]	Size included? (assessment)[e]
Weaver and Cavers (1980)	Rumex crispus – P	4	Age (fields, waste areas)	Yes	Yes (no effect)
Weaver and Cavers (1980)	Rumex obtusifolius – P	4	Ph (open to forest)	Yes	Yes (no effect)
Grace and Wetzel (1981)	Typha latifolia – P	3	Ph (open to forest)	Yes	No
Soule and Werner (1981)	Potentilla recta – P	3	Ph + age (open to forest)	Right trend	No (positive sd RA?)
Cartica and Quinn (1982)	Solidago sempervirens – P	4	Ph (open to forest)	Yes	Yes (influential?)
Primack and Antonovics (1982)	Plantago lanceolata – P	6–8	Ph (all open)	Yes	No
Stewart and Thompson (1982)	Carex flacca – P	3	Ph (open to bushy)	No	No
Stewart and Thompson (1982)	Centaurea nigra – P	3	Ph (open to bushy)	Right trend	No
Stewart and Thompson (1982)	Leontodon hispidus – P	3	Ph (open to bushy)	No	No
Stewart and Thompson (1982)	Plantago lanceolata – P	3	Ph (open to bushy)	No	No
Stewart and Thompson (1982)	Plantago media – P	3	Ph (open to bushy)	Right trend	No
Stewart and Thompson (1982)	Poterium sanguisorba – P	3	Ph (open to bushy)	No	No
Waite and Hutchings (1982)	Plantago coronopus – A	3	Ph (dune, not clear)	? (1 odd)	Only mentioned
Jurik (1983)	Fragaria virginiana – P	3–4	Ph (open to bushy)	Yes, 1976	No

Jurik (1983)	*Fragaria virginiana* – P	3–4	Ph (open to bushy)	?, 1 odd 1977	No
Studies limited to the same genus					
Abrahamson and Gadgil (1973)	4 *Solidago* – P	3	Ph (open to forest)	Yes	No – Yes (no influence)
Gaines *et al.* (1974)	4 *Helianthus* – 1 A, 3 P	4	Ph (open to forest)	? (1 odd)	Yes, no trend if annual sp. excluded – No
Brouillet and Simon (1979)	2 *Aster* – P	2 (rep)	Ph (open to forest)	No	No – Yes (sd RA)
Harrison (1979)	2 *Zostera* – P	1 (2 levels)	Tidal level	Yes	No – Yes (sd RA in one species)
Primack *et al.* (1981)	5 *Solidago* – P	3	Ph (open to forest)	Yes	No – Yes (sd RA)
Gross *et al.* (1983)	3 *Solidago* – P	2	Age (open to forest)	Yes	No – Yes (si RA)
Gross *et al.* (1983)	2 *Aster* – P	2	Age (open to forest)	Yes	No – Yes (si RA)
Jurik (1983)	2 *Fragaria* – P	2	Ph (open to forest)	Yes	No – No
Whole community studies					
Struik (1965)		2 (rep)	Ph (open versus forest)	No	Annuals and perennials studied separately
Gadgil and Solbrig (1972)		3	Ph + dis (open to forest)	? (1 odd)	Cannot be assessed
Newell and Tramer (1978)		3	Age (open to forest)	? (asymp)	Likely (turnover stated by authors)
Abrahamson (1979)		2 (rep)	Age (open versus forest)	Yes	Yes
Luftensteiner (1980)		4	Ph (open to forest)	No	Very few annual species
Cid-Benevento and Werner (1986)		2*	Ph (open to forest)	No	Only annual species included – No (sd RA)

[a] A: annual; P: perennial; A–P: annual to perennial.
[b] rep: replicates within each kind of habitat, are indicated within brackets.
[c] dis: disturbance; Ph: physiographic (vegetation) description.
[d] Right trend: trend in the direction predicted, but nonsignificant; asymp: asymptotic trend; 1 odd: one population destroys the trend.
[e] sd: size-dependent; si: size-independent.
*Greenhouse study, 2 origins.

Appendix 3

Data Sets Testing the Prediction of Decreasing RA with Increased Density (Competition) and Decreased RA with Decreasing Disturbance

Reference	Species – life history	Design	Treatments	Support?
Density				
Ogden (1974)	*Tussilago farfara* – P	Greenhouse	4	Yes
Abrahamson (1975b)	*Rubus hispidus* – P	Field	2	RE increased with density
Abrahamson (1975b)	*Rubus trivialis* – P	Field	2	No
Snell and Burch (1975)	*Chamaesyce hirta* – P	Greenhouse	4	Yes
Holler and Abrahamson (1977)	*Fragaria virginiana* – P	Greenhouse	2	No
Hawthorn and Cavers (1978)	*Plantago major* – P	Greenhouse	3	No
Hawthorn and Cavers (1978)	*Plantago rugelii* – P	Greenhouse	3	Yes? (total suppression of flowering)
Abul-Fatih *et al.* (1979)	*Ambrosia trifida* – A	Greenhouse	5	Yes
Pitelka *et al.* (1980)	*Aster acuminatus* – P	Field	7	No
Winn and Pitelka (1981)	*Aster acuminatus* – P	Field	5	No
Waite and Hutchings (1982)	*Plantago coronopus* – A	Greenhouse	3	Yes
Ashmun *et al.* (1985)	*Aster acuminatus* – P	Field	2	Yes/No
Augspurger *et al.* (1985)	*Sesbania emerus* – P	Field	3	Yes (controlled for size)

				Yes/No
Carpenter and West (1988)	*Artemisia tridentata* ssp. *vaseyana* – P (shrub)	Field	2	Yes/No
Clauss and Aarssen (1994)	*Arabidopsis thaliana* – A	Greenhouse	6	No (controlled for size)
DeFalco *et al.* (2003)	*Bromus matritensis* ssp. *rubens* – A	Greenhouse	2	Only at high nutrients
DeFalco *et al.* (2003)	*Vulpia octoflora* – A	Greenhouse	2	Increasing at low nutrients; decreasing at high nutrients
DeFalco *et al.* (2003)	*Descurainia pinnata* – A	Greenhouse	2	No
Disturbance				
Dunn and Sharitz (1991)	*Murdannia keisak* – A	Field	4	Yes
Dunn and Sharitz (1991)	*Murdannia keisak* – A	Greenhouse	2	No
Kobayashi *et al.* (1999)	*Plantago asiatica* – P	Common garden	2	Opposite response
Kobayashi *et al.* (1999)	*Artemisia princeps* – P	Common garden	2	Reproduction impaired under disturbance
Kobayashi *et al.* (1999)	*Eleusine indica* – A	Common garden	2	No
Kobayashi *et al.* (1999)	*Digitaria adscendens* – A	Common garden	2	No

Appendix 4

Summary of Studies Testing the Response of Annual and Perennial Plants to Different Stress Treatments

Reference	Species – life history	Design	Treatment[*]	Response to stress
Nutrient Stress				
Harper and Ogden (1970)	*Senecio vulgaris* – A	Greenhouse	Pot size (3)	?[a] (also smaller size)
Snell and Burch (1975)	*Chamaesyce hirta* – A	Greenhouse	Nutrients (4)	Decreased RA
van Andel and Vera (1977)	*Senecio sylvaticus* – A	Greenhouse	Nutrients (3)	No response
Weiss (1978)	*Emex australis* – A	Greenhouse	Pot size (3)	Decreased RA (also smaller size)
Waite and Hutchings (1982)	*Plantago coronopus* – A	Greenhouse	Nutrients (3)	No response
Fenner (1986)	*Senecio vulgaris* – A	Greenhouse	Nutrients (5)	No response (smaller size)
Cheplick (1989)	*Amphicarpum purshii* – A	Greenhouse	Nutrients (2)	Decreased RA (no size effect)
Schat et al. (1989)	*Centaurium pulchellum* – A	Field	Nutrients (2)	Decreased RA (no size effect)
Benech Arnold et al. (1992)	*Sorghum bicolor* – A	Greenhouse	Nutrients (3)	Decreased RA (also smaller size)
Hermanutz and Weaver (1996)	*Solanum ptycanthum* – A	Greenhouse	Nutrients (3)	Increased RA (also smaller size)
DeFalco et al. (2003)	*Bromus madritensis* ssp. *rubens* – A	Greenhouse	Nutrients (2)	Increased RA
DeFalco et al. (2003)	*Vulpia octoflora* – A	Greenhouse	Nutrients (2)	No response at low density; Increased RA at high density
DeFalco et al. (2003)	*Descurainia pinnata* – A	Greenhouse	Nutrients (2)	No response
Lovett Doust (1980b)	*Smyrnium olusatrum* – B	Greenhouse	Nutrients (2)	Decreased RA (also smaller size)
Hansson (1996)	*Anthriscus sylvestris* – B	Greenhouse	Nutrients (2)	Decreased RA
Ogden (1974)	*Tussilago farfara* – P	Greenhouse	Nutrients (2)	No response
van Andel and Vera (1977)	*Chamaenerium angustifolium* – P	Greenhouse	Nutrients (3)	No response

Bostock (1980)	*Tussilago farfara* – P	Field	Nutrients (3)	Decreased RA
Primack and Antonovics (1982)	*Plantago lanceolata* – P	Greenhouse	Nutrients (2)	Decreased RA
Pratt (1984)	*Solidago juncea* – P	Field	Nutrients (4)	Decreased RA (also smaller size)
Pratt (1984)	*Solidago graminea* – P	Field	Nutrients (4)	Decreased RA (one year later)
Alpert *et al.* (1985)	*Diplacus aurantiacus* – P (bush)	Field	Nutrients (2)	No response (smaller size)
Boutin and Morisset (1988)	*Chrysanthemum leucanthemum* – P	Greenhouse	Nutrients (3)	No response (smaller size)
Carpenter and West (1988)	*Artemisia tridentata* ssp. *vaseyana* – P (bush)	Field	Nutrients (2)	Increased RA (1st year), no response (2nd year)
Benech Arnold *et al.* (1992)	*Sorghum halepense* – P	Greenhouse	Nutrients (3)	No response (smaller size)
Cheplick (1995)	*Amphibromus scabrivalvis* – P	Greenhouse	Nutrients (2)	Decreased RA/No response
Water stress				
Hickman and Pitelka (1975)	*Polygonum cascadense* – A	Field	Moisture (2)	Increased RA (also smaller size)
Abrahamson and Hershey (1977)	*Impatiens capensis* – A	Field	Moisture (2)	No response (also smaller size)
Abrahamson and Hershey (1977)	*Impatiens capensis* – A	Greenhouse	Moisture (2)	No response (similar size)
Hickman (1977)	*Polygonum douglasii* – A	Field	Moisture (3)	No consistent trend
Hickman (1977)	*Polygonum kelloggii* – A	Field	Moisture (3)	Highest RA in wettest
Hickman (1977)	*Polygonum minimum* – A	Field	Moisture (2)	No response
Jaksic and Montenegro (1979)	*Chaetanthera ciliata* – A	Field	Moisture (2)	Higher RA in dry year
Jaksic and Montenegro (1979)	*Erodium cicutarium* – A	Field	Moisture (2)	Higher RA in dry year
Jaksic and Montenegro (1979)	*Trisetobromus hirtus* – A	Field	Moisture (2)	Higher RA in dry year
Cid-Benevento and Werner (1986)	*Chenopodium album* – A	Greenhouse	Moisture (3)	No response/No consistent trend
Cid-Benevento and Werner (1986)	*Polygonum pensylvanicum* – A	Greenhouse	Moisture (3)	No consistent trend
Cid-Benevento and Werner (1986)	*Acalypha rhomboidea* – A	Greenhouse	Moisture (3)	No response

(Continued)

Appendix 4

Summary of Studies Testing the Response of Annual and Perennial Plants to Different Stress Treatments—cont'd

Reference	Species – life history	Design	Treatment*	Response to stress
Cid-Benevento and Werner (1986)	*Pilea pumila* – A	Greenhouse	Moisture (3)	Decreased RA (at lower light)
Cid-Benevento and Werner (1986)	*Impatiens capensis* – A	Greenhouse	Moisture (3)	No response
Sobrado and Turner (1986)	*Helianthus petiolaris* – A	Greenhouse	Moisture (2)	No response
Aronson *et al.* (1993)	*Brachypodium distachyon* – A	Greenhouse	Moisture (2)	No response
Aronson *et al.* (1993)	*Bromus fasciculatus* – A	Greenhouse	Moisture (2)	Increased RA
Aronson *et al.* (1993)	*Erucaria hispanica* – A	Greenhouse	Moisture (2)	Increased RA
Jaksic and Montenegro (1979)	*Solenomelus peduncularis* – P	Field	Moisture (2)	Same RA both years
Alpert *et al.* (1985)	*Diplacus aurantiacus* – P (bush)	Field	Moisture (2)	No response
Augspurger *et al.* (1985)	*Sesbania emerus* – P	Field	Moisture (3)	Increased RA (controlled for plant size)
Carpenter and West (1988)	*Artemisia tridentata* ssp. *vaseyana* – P (bush)	Field	Moisture (2)	Increased RA (1st year), no response or lower RA (2nd year)
Grulke and Bliss (1988)	*Phippsia algida* – P	Field	Moisture (2)	Lower RA in drier year
Grulke and Bliss (1988)	*Puccinellia vaginata* – P	Field	Moisture (2)	Lower RA in drier year
Li *et al.* (2001b)	*Cyperus esculentus* – P	Greenhouse	Moisture (2)	Increased RA (also smaller size)
Light stress				
Cid-Benevento and Werner (1988)	*Chenopodium album* – A	Greenhouse	Light (6)	Decreased RA (also smaller size)
Cid-Benevento and Werner (1988)	*Polygonum pensylvanicum* – A	Greenhouse	Light (6)	No consistent trend (smaller size)
Cid-Benevento and Werner (1988)	*Acalypha rhomboidea* – A	Greenhouse	Light (6)	Decreased RA at lowest light (also smaller size)

Reference	Species	Environment	Treatment	Result
Cid-Benevento and Werner (1988)	*Pilea pumila* – A	Greenhouse	Light (6)	Maximum RA at intermediate light levels (decreasing size)
Cid-Benevento and Werner (1988)	*Impatiens capensis* – A	Greenhouse	Light (6)	Increased RA (decreasing size)
Hansson (1996)	*Anthryscus sylvestris* – B	Greenhouse	Light (2)	Decreased RA
Roos and Quinn (1977)	*Andropogon scoparius* – P	Greenhouse	Light (2)	Decreased RA (similar size)
Pitelka *et al.* (1980)	*Aster acuminatus* – P	Field	Light (7)	Decreased RA (also smaller size)
Young (1983)	*Arnica cordifolia* – P	Field	Light (3)	Decreased RA (increasing size)
Alpert *et al.* (1985)	*Diplacus aurantiacus* – P (bush)	Field	Light (2)	Decreased RA (also bigger size)
Ashmun *et al.* (1985)	*Aster acuminatus* – P	Field	Light (2)	Decreased RA (also smaller size)
Ågren (1988)	*Rubus chamaemorus* – P	Field	Light (2)	Decreased RA
Boutin and Morisset (1988)	*Chrysanthemum leucanthemum* – P	Greenhouse	Light (3)	Increased RA
Verburg and Grava (1998)	*Circaea lutetiana* – P	Greenhouse	Light (2)	No response (controlled for plant size)
Li *et al.* (2001a)	*Cyperus sculentus* – P	Greenhouse	Light (4)	Decreased RA (also smaller size)
Altitude				
Kawano and Masuda (1980)	*Heloniopsis orientalis* – P	Field	Altitude (5)	Increased RA at flowering[b]
Kawano and Masuda (1980)	*Heloniopsis orientalis* – P	Field	Altitude (5)	Increased RA at fruiting
Douglas (1981)	*Mimulus primuloides* – P	Field	Altitude (5)	No pattern
Jolls (1984)	*Sedum lanceolatum* – P	Field	Altitude (4)	Decreased RA[c]

[a] No difference between low and medium stress treatments (trend in the right direction), reproduction suppressed in the high-stress treatment.
[b] Marginally significant.
[c] Opposite trend when using energy as currency.
* Number of treatment levels within brackets.

2

Meristem Allocation as a Means of Assessing Reproductive Allocation

Kari Lehtilä, Annika Sundås Larsson

I. Introduction

The principle of resource allocation states that resources are not sufficient to fully supply the demands of all plant functions (Gadgil and Bossert, 1970). If resources are limited, allocation to reproduction must decrease the allocation to some other plant functions. Allocation of resources to reproduction (reproductive allocation) has implications for several study fields in biology. Early studies discussed allocation patterns in light of the evolution of life-history traits (Williams, 1966; Gadgil and Bossert, 1970) and interaction between life-history traits and population dynamics (r and K selection; MacArthur and Wilson, 1967). Initial studies considered animals, but the idea was soon applied to plant ecology to explain how allocation affects plant life-history strategies and plant community dynamics (Grime, 1979).

To test the principle of allocation, biologists have tried to measure the resource use of different plant functions. This task has turned out to be difficult. It is not clear which currency should be used to measure reproductive allocation. Biomass, energy, carbon, or mineral nutrients have been used as a currency of allocation (Reekie, 1999). Furthermore, it is not easy to delimit plant functions to specific parts of a plant. For instance, allocation to reproduction may be defined as allocation to seeds or fruits; all other plant parts are in that case regarded as vegetative parts and allocation to them can be called as somatic allocation. But production of flowers and structures supporting reproductive parts also have costs, although they may compensate carbon needs by their own photosynthesis (Bazzaz *et al.*, 1979). In an attempt to solve these problems, meristems have been proposed as a currency to measure reproductive allocation (Watson, 1984; Geber, 1990). Meristems come as distinct reproductive or vegetative entities. Reproductive allocation can then be defined as the fraction of inflorescence and floral meristems, and somatic allocation as the fraction of other meristem types out of all meristems of a plant. If the estimation of proportions of reproductive and vegetative parts is found to be a suitable method to measure reproductive allocation, it may be more straightforward to count meristems than to delimit reproductive and vegetative parts in some other way.

There are two reasons why meristem allocation may be a good measure of reproductive allocation. First, meristem allocation may work as an indicator of underlying allocation patterns of resources. Meristems may be more practical to use than other currencies, and they may work as a surrogate of critical resources, because a plant that produces many flowers or inflorescences usually has a higher allocation to reproduction than a plant that devotes a few meristems to reproduction. To test whether using meristems produces the same results as that using some other currencies in the measurement of resource allocation, the suitability of alternative currencies must be validated, at least in some cases. Second, meristem allocation may itself be a process that molds patterns of reproduction (Watson, 1984; Geber, 1990). Meristem availability may constrain possibilities to invest in reproduction, at least in some environments (Geber, 1990). In this way, meristems may limit reproduction instead of, or in addition to, essential resources. On the other hand, meristem production is ultimately dependent on resource levels. It has been suggested that meristem limitation is more likely when resource levels are good, so that plants with few meristems are not able to invest all available resources to growth and reproduction (Geber, 1990).

Our aim is to discuss whether meristem allocation can be used as a supplement or a substitute when studying reproductive allocation. We will first describe the structural, developmental, and physiological background of the meristem system. We then use a modeling approach to show theoretically how meristem allocation may affect fitness. We discuss model assumptions and review empirical studies that have tested allocation models and their assumptions.

II. Developmental and Physiological Background of Meristem Allocation

After the embryonic stage, most plant cell divisions take place in meristems, tissues with undifferentiated cells. The meristem type that is the target of this chapter is the shoot apical meristem (SAM), which is formed during embryogenesis. During growth, SAMs produce leaves and new meristems in leaf axils. Axillary meristems are also SAMs. In addition to SAMs, plants have other meristem types, such as root apical meristems, lateral meristems, and meristematic tissue contributing to leaf growth. Some plant groups have special meristem types such as intercalary meristems of grasses. In plants, where organs generate from meristems throughout the life cycle, environmental factors affect the growth and development much more than in animals. Thus, plant development is a process regulated by the endogenous genetic program as well as external environmental signals. The architecture of the plant is determined by the number of meristems produced as well as the positions of the meristems. The following three sections will describe the structure, development, and function of the apical meristem in terms of gene regulation and physiology affecting the initiation and subsequent development of the meristem. In recent years, the analysis of the model plant *Arabidopsis thaliana* (Brassicaceae) has lead to a large amount of knowledge concerning plant meristems and their regulation.

III. Meristem Structure and Generation of Plant Architecture

The meristem has been described as either a layered structure or as composed of different zones. Strict regulation of the plane of cell division gives rise to the layered appearance, and the division of the meristem into zones is based on the rate of cell division of the central zone (site of meristem maintenance) when compared to the peripheral zone where primordia are initiated (Fig. 2.1). As it turns out, these descriptions should be superimposed to better describe the highly organized structure of the meristem. The meristem organization, the initiation of organ primordia, and models of the genetic regulation of the maintenance of the structure has been reviewed by Traas and Vernoux (2000), Brand *et al.* (2001), and Fletcher (2002).

The activity of the meristem described above is strictly regulated in the sense that a plant is always built according to a genetically determined pattern, the phyllotaxy, which poses a constraint on the developmental plasticity (Napoli *et al.*, 1999). The basic building unit of plants is the phytomer composed of the internode (a stem segment), the node (the point where a leaf/leaves are attached), and a meristem in the leaf axil. The number of leaves at each node and the angle between sequentially initiated leaves are species-specific characters. The axillary meristem gives rise to a bud which can be dormant or can develop into a shoot. In addition, some plants produce one or two accessory buds near the axillary bud. Thus, the plant architecture, which determines the production of new meristems, is regulated at the genetic level. However, environmental conditions can still affect meristem production and, to a greater extent, meristem identity. Furthermore, meristem identity affects subsequent meristem allocation. Some "growth habits" restrict the plasticity of meristem allocation, e.g., once monopodial annuals go through floral transition, they lose the ability to produce axillary meristems that give rise to new inflorescences. Both phyllotaxy and floral transition are factors that may highly affect the pattern of meristem initiation and allocation.

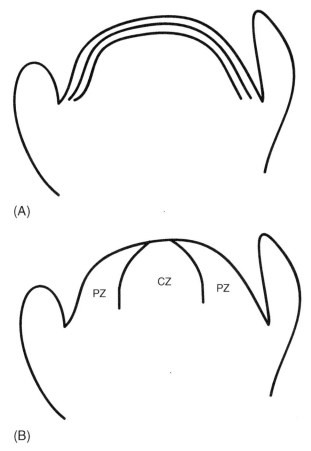

Figure 2.1   The histology of the shoot apical meristem. (A) The SAM is composed of clonally distinct cell layers. (B) In the SAM, the peripheral zone (PZ) and the central zone (CZ) can be distinguished based on cell division rates.

Do plants develop shoot meristems and branches other than the ones described above? Plants can develop so-called adventitious buds, which are not situated in axillary positions. Meristem allocation studies, however, usually regard the contribution of adventitious buds on meristem dynamics as negligible. Obviously, this assumption has to be confirmed in each of the empirical studies. Further anomalies are lateral branches that are formed from stems (epicormic), roots (epirhizic), and leaves (epiphyllous) (Kerstetter and Hake, 1997). None of these meristems/branches normally make any major contribution to the plant shoot system of most plant species and can usually be neglected if meristem allocation is used to assess reproductive allocation.

IV. Axillary Bud Formation and Subsequent Development of the Bud

In most species, axillary meristems do not form in contact with the SAM itself but in the leaf axil. The development of axillary meristems has been described as originating either from cells that have retained the meristematic properties from the SAM (the detached meristem concept of axillary meristem origin) (Garrison, 1949a,b, 1955; Remphrey and Steeves, 1984; Wardlaw, 1943) or from cells that have regained such properties (Majumdar, 1942; Sharman, 1942). Even though studies in Arabidopsis support the former concept (Grbic and Bleeker, 2000; Long and Barton, 2000), it cannot be excluded that they can arise from cells that regain meristematic properties.

The architecture of the plant is highly dependent on whether the axillary meristems develop after inititation. Axillary meristems in axils of older leaves develop during an extended period of vegetative growth at a distance from the apex. After the transition to flowering, two things change: first, the pattern of axillary meristem development is reversed and the meristems in axils of the youngest leaves are initiated to grow, which produce a branched inflorescence; second, the SAM gives rise to floral meristems that develop into flowers rather than shoot meristems. Axillary shoot meristems that are already initiated can, however, develop into flowers when subjected to strong photoinduction treatment (Hempel *et al.,* 1998). This shows that the specification of the axillary bud is a progressive process rather than a fixed acquired fate and that plant development is highly plastic and affected by environmental signals.

There is no clear picture on the physiological regulation of the meristem initiation and development, but several studies show that the plant hormones auxin, cytokinin, and abscisic acid are involved in the processes. Apical dominance is a well-known concept in the control of bud outgrowth, which states that the apical growth point controls the outgrowth of axillary buds. The involvement of the plant hormone auxin in apical dominance has long since been recognized (Tamas, 1995). In *Lupinus angustifolius* (Fabaceae), the relative amounts of cytokinin and auxin regulate axillary bud development during the early stages and another hormone, abscisic acid, correlates with bud development during the later stages (Emery *et al.,* 1998; Miguel *et al.,* 1998).

Mutations in the Arabidopsis gene *SUPERSHOOT* (*SPS*) show defects in the process of axillary meristem initiation (Tantikanjana *et al.,* 2001). The normal pattern of one axillary meristem in each leaf axil is perturbed in the mutant in which several meristems are formed in each axil, and further, they all develop to give rise to plants with 500 or more inflorescences. The results suggest that *SPS* has a role in suppressing meristem initiation and growth by decreasing cytokinin levels in the leaf axils. The evolving picture of regulation of branch development shows that it is determined by the ratio of several of the known plant hormone levels, which can change during development. A number of other Arabidopsis mutants with perturbances in branching pattern and hormone levels exist and should in time shed light on this process.

Apart from hormone physiology, environmental factors such as light quality and nutrition also have the potential to affect meristem allocation. Several species, including Arabidopsis, have a shade avoidance response whereby shading increases apical dominance (fewer buds develop into branches) and results in internode elongation/petiole elongation and reduced leaf area. In addition, flowering is accelerated and there is a reduction of allocation of resources for storage and reproduction (Smith, 1995; Smith and Whitelam, 1997; Morelli and Ruberti, 2000).

V. Genetics and Physiology of the Floral Transition

In Arabidopsis as well as other monopodial annuals, the architecture of the plant is highly dependent on when the transition to flowering takes place. After the transition, mainly floral meristems are initiated. Flowering time is closely linked to the reproductive success of the plant and is regulated by a complex genetic network involving genes in the endogenous developmental program and in the response to external signals. More than 80 genes are known to be involved in the regulation of flowering time in Arabidopsis (reviewed by Simpson *et al.*, 1999), and four interconnected pathways have been identified by genetic analysis of flowering time mutants. The autonomous pathway promotes flowering independent of environmental conditions, monitoring endogenous cues, such as age. The pathway involving hormonal control, the GA pathway, promotes flowering in noninductive short-day conditions. Two other pathways mediate the plant response to external cues, the long-day pathway and the vernalization pathway, mediating photoperiodic and low temperature responses, respectively. Ultimately, these pathways converge on the regulation of a small number of genes that activate the floral organ identity genes in meristems determined to develop as flowers. One of these genes, *LEAFY* (*LFY*), has been shown to integrate much of the information regarding floral induction (Blázquez, 2000).

Further, nutrition also has an impact on flowering time. It is known that sucrose has an effect in Arabidopsis. High levels of sucrose cause a delay of the transition to the reproductive phase, allowing for more meristems to be allocated to further growth (Ohto *et al.*, 2001). In most cases, different types of stress shorten the time to flowering. It is not known till date whether these factors act through one of the described pathways or if there are additional unidentified pathways.

VI. Meristem Types

For modeling and empirical field studies, the different types of shoot apical meristems are often simplified in main categories. There are several alternative ways of classification. As a starting point, meristem states can be presented as follows:

(1) *Dormant buds* have not started their growth after initiation in axillary positions. It should be noted that due to their properties, even dormant buds are regarded as shoot apical meristems, although they do not have an apical position before growth.

(2) Active meristems:

    (2a) *Vegetative meristems* have continuous growth and they are able to produce axillary meristems. Vegetative meristems are responsible for growth, branching, and expansion of the shoot system.

    (2b) *Inflorescence meristems* have continuous growth similar to vegetative meristems, but they mainly produce floral meristems (state (2c)) in axillary positions. Many plant species that produce a terminal flower do not have inflorescence meristems, or the inflorescence stage is a very brief period before transformation to floral meristems.

    (2c) *Floral meristems* are terminal and produce one flower, i.e., the meristem is consumed when the flower is formed.

(3) *Senescent, inactive meristems,* the meristems of states (1), (2a), and (2b) finally senesce and become inactive, as also occasionally meristems of state (2c).

Ecological literature often uses meristem death as one of the meristem states but in most cases inactive meristems contain living cells, although no cell proliferation occurs as resources are allocated elsewhere in the plant. Meristems senesce with the rest of the plant, and cell death occurs only if the whole plant dies or if a plant part and its meristems die and are shed.

Essential for meristem allocation is that floral meristems are terminal and that reproductive meristems (2b) and (2c) usually do not transform back to vegetative meristems. The transformations to inflorescence and floral meristems thus decrease the number of meristems of other types. The principle of allocation, adapted to meristems, states that high allocation to inflorescence and floral meristems decreases the number of vegetative meristems. Reproductive allocation may then have a cost through reduced future production of new meristems and expansion of the shoot system.

This model of the meristem system may be complemented with description of dynamics. Meristem dynamics consists of:

(A) The schedule of production of new buds by vegetative and inflorescence meristems (2a) and (2b). It is assumed that new buds enter the dormant state (1) before transformation to other states.

(B) The schedule of activation of dormant buds, which may then enter any of the active stages (2a), (2b), or (2c).

(C) The schedule of transformation from vegetative state (2a) into inflorescence meristem (2b). Vegetative meristems (2a) or inflorescence meristems (2b) do not transform into floral meristems (2c). The exception is when the (2a) or (2b) occur as a short transient state in development from dormant buds to floral meristems.

(D) Bud senescence can be implicit in the dynamics, or the schedule may explicitly state when bud production ceases and the meristem enters state (3).

Bonser and Aarssen (1996) presented a simpler model of the meristem system which is based on three meristem types. In their model, inactive (I) meristems are dormant buds or senescent meristems. Growth (G) meristems are equal to the state (2a) (vegetative meristems) and inflorescence and floral meristems are pooled as reproductive (R) meristems. Bonser and Aarssen (1996) define branch tips as inactive even when branches are growing actively, which means that maximally 50% of the meristems can be G meristems. Other definitions are also possible, such as considering actively growing vegetative apices as G meristems, which become I meristems only when they senesce (Huber and During, 2001).

Based on species characteristics, other classifications of meristems may be relevant. Since meristem state and fate varies among meristems of basal and apical location or in the sun and shade, the meristem position within the canopy can be included in the analysis (Jones, 1985). Heterophyllous trees such as birches have two types of shoots (Macdonald and Mothersill, 1983; Macdonald *et al.*, 1984). Long shoots have sympodial growth. They start growing from axillary meristems of the previous year, and produce leaves and axillary buds throughout the growing season until the terminal bud becomes inactive at the end of the season or produces a flower. Short shoots have very short internodes and produce their leaves in a single burst at the beginning of the season. Short shoots are monopodial, i.e., the terminal bud continues its growth for many years until it senesces or produces a flower. In analyzing meristem systems of heterophyllous plants, short and long shoots must be kept separate, because they differ in their dynamical properties (Lehtilä *et al.*, 1994).

VII. Meristem Models

We can evaluate by means of mathematical models how meristem allocation patterns affect individual fitness. Development and structure of meristem systems produce regular patterns that make them amenable to modeling. Meristem systems resemble structured populations, where different meristem states correspond to different types of individuals. Models derived from population dynamics, such as population matrix models, have thus often been used in meristem allocation studies (Maillette, 1982; Lehtilä *et al.*, 1994). Huber and During (2001) constructed a simulation model of meristem production of *Trifolium* sp. (Fabaceae), in which growth meristems produced axillary buds that either immediately became flowers or, alternatively, remained dormant and became growth meristems after the mother meristem had produced two more buds. Huber and During (2001) varied the fraction of axillary buds that become flowers. After a short season of ten time steps (meristem production cycles), the maximal fecundity was attained when 90–100% of meristems were allocated to flowering. This means that if the growing season is very short and plants (or aboveground shoots) die at the end of the season, sexual reproduction is maximized with a high allocation to flowering. After twenty and thirty time steps, the optimal allocation to flowering was 40 and 30%, respectively. A longer growing season favors a strategy with lower reproductive allocation. The model suggests that reproductive allocation depends on the length of the season.

It is possible to use an analytical approach to the problem. If a plant without reproduction has N_{t-1} growing meristems at time $t-1$ and each meristem contributes to the meristem number at the next time step by, on average, n meristems, the number of meristems at time t will be $N_t = nN_{t-1}$. In reproductive plants, a fraction r ($\leq n$) of the meristems are allocated to reproduction. In this case, $N_t = (n - r)N_{t-1}$. If one reproductive meristem produces k seeds, the reproductive output at time t is $R_t = krN_t$. The total fitness up to time t, W_t, can be calculated as a sum of R_is where $i = 1 \dots t$. The coefficient n includes both the average survival of the meristems and production of new buds. To simplify the model, it is presently assumed that the production of dormant buds has no notable contribution to bud dynamics other than to reduce n. Other possible functions of dormant buds are discussed later.

Assuming that rates of meristem production and flowering remain constant over time, similarly as in the model of Huber and During (2001), meristem number can be calculated iteratively backwards as $N_t = (n - r)N_{t-1} = (n - r)(n - r)N_{t-2}$ and so on. Iteration produces the following equations for meristem number, reproductive output, and fitness at time t:

$$N_t = (n - r)N_{t-1} = (n - r)^t N_0 \tag{2.1}$$

$$R_t = krN_{t-1} = kr(n - r)^{t-1} N_0 \tag{2.2}$$

$$W_t = \sum_{i=1}^{t} R_i = \sum_{i=1}^{t} kr(n - r)^{i-1} N_0 \tag{2.3}$$

N_0 is the number of buds at the time point early in the season, when rates of meristem production and flowering have stabilized. Growth starts, in principle, from one meristem, but plants usually have a vegetative growth period before they attain the threshold size of flowering. Therefore, N_0 is in that case the number of meristems at the onset of flowering rather than the initial meristem.

The optimal meristem allocation to reproduction is calculated by finding parameter values where $\partial W_t / \partial r = 0$. An additional potential optimal value is $r = n$, i.e., immediate allocation of all meristems to reproduction. Derivative of W_t with respect to r is

$$\frac{\partial W_t}{\partial r} = \sum_{i=1}^{t} \frac{\partial R_i}{\partial r} = \sum_{i=1}^{t} k(n-r)^{i-2}\left[n - r - (i-1)r\right]N_0 \qquad (2.4)$$

The optimal resource allocation depends on the specific values of n and N_0, but the inspection of the $\partial R_i / \partial r$ values within summations reveal that there is often an intermediate optimal value of r (Fig. 2.2). The $\partial R_i / \partial r$ values

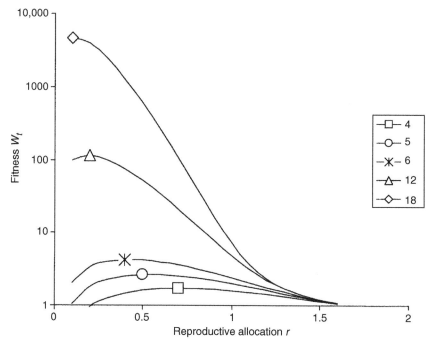

Figure 2.2   Effect of allocation to reproductive meristems (r) on fitness (W_t) (Eq. 2.3 in the text). Lines represent fitness as a function of reproductive allocation for different season lengths, measured as time steps of meristem production cycles ($t = 4 \ldots 18$). Optimal reproductive allocations are denoted with symbols. Growth rate of meristem system $n = 2$. Meristem number at the onset of flowering $N_0 = 4$. Number of seeds produced per reproductive meristem $k = 0.1$.

are larger than zero when $i < n/r$. When reproductive allocation r is a very small fraction of meristem production n, the condition $i < n/r$ is true for all values of i, even when $i = t$. Since all $\partial R_i/\partial r$ values are larger than zero, $\partial W_t/\partial r$ is larger than zero. Fitness increases then with increased r, i.e., there is a selection pressure for higher values of r. When reproductive allocation is large, the $\partial R_i/\partial r$ values may be negative even for small values of i. For instance, if half of the meristems are allocated to reproduction, $\partial R_i/\partial r$ becomes negative at the second time step $(i = 2)$. It is then probable that there is a selection pressure for reduced values of r. In summary, there is an intermediate critical level of reproductive allocation r that is at least local optimum. If the number of meristem cycles before plant death increases, the optimal reproductive allocation decreases and vice versa.

One potential critical point of reproductive allocation is $r = n$, which represents the "bang-bang reproduction," immediate allocation of all resources to reproduction at the onset of flowering (Cohen, 1971). At that point, fitness is $W_t = R_1 = knN_0$ for all values of t. It is possible to continue the analysis by comparing this fitness value with fitness values of intermediate optimal r, which would test whether the winning strategy is a bang-bang reproduction or continuous reproduction over a longer time.

One can also change the assumptions of the model and ask what the optimal outcome would be when reproductive allocation is not constant over time. Would it be beneficial to have different allocation patterns at different points in time?

Adding plasticity of reproductive allocation to the assumptions of the model of Eqs (2.1)–(2.4), the optimal strategy is to allocate the meristems to growth until the last time step and then to allocate all meristems to reproduction. We modify the model by recording the reproductive allocation as the number F_t of meristems devoted to reproduction at time t, not as a fraction r. In this case, $N_t = n(N_{t-1} - F_{t-1}) = n(n(N_{t-2} - F_{t-2}) - F_{t-1})$, which results in

$$N_t = n^t N_0 - \sum_{i=0}^{t-1} n^{t-i} F_i \qquad (2.5)$$

Fitness at time t is

$$W_t = k \sum_{i=0}^{t-1} F_i \qquad (2.6)$$

According to this model, the best strategy is bang-bang reproduction. A convenient method to show this is to consider two strategies, denoted with subscripts A and B, and evaluate their fitness at $t + 1$, the time of death. It can be shown that fitness is always greater for the strategy with lower reproductive allocation. We assign that A and B have the same reproductive allocation, $F_{iA} = F_{iB}$, except for one time step j, when reproductive allocation is one meristem larger for strategy A, i.e., $F_{jA} = F_{jB} + 1$. The number of meristems at time t for strategy A, N_{tA}, can be expressed with N_{tB} by using Eq. (2.5) and noting that F values are the same except for time step j:

$$N_{tA} = n^t N_0 - \sum_{i=0}^{t-1} n^{t-i} F_{iA} = n^t N_0 - \sum_{i=0}^{j-1} n^{t-i} F_{iA} - n^j F_{jA} - \sum_{i=j+1}^{t-1} n^{t-i} F_{iA}$$

$$= n^t N_0 - \sum_{i=0}^{j-1} n^{t-i} F_{iB} - n^j \left(F_{jB} + 1 \right) - \sum_{i=j+1}^{t-1} n^{t-i} F_{iB} = N_{tB} - n^j \quad (2.7)$$

This result shows that the number of meristems before the last time step is lower for strategy A than for strategy B. On the other hand, realized fitness is higher for strategy A at that time step. Realized fitnesses before the last time step are $W_{tA} = W_{tB} + k(F_{jA} - F_{jB}) = W_{tB} + k$, because strategy A produced one more reproductive meristem than strategy B. It is evidently beneficial to allocate all remaining meristems to reproduction at the time of death, so that $F_t = N_t$ and $W_{t+1} = W_t + k N_t$. Thus, $W_{t+1,A} = W_{tB} + k + k(N_{tB} - n^j) = W_{t+1,B} - (n^j - 1)k$. Strategy B, with lower reproductive allocation, has higher fitness. By induction we can conclude that fitness of any strategy can be increased by reducing reproductive allocation at time steps $t = 0$ ... $t-1$, until reproductive allocation is zero for all except the last time step.

VIII. The Assumptions of the Models

The first model (Eqs (2.1)–(2.4)) and the model of Huber and During (2001) predicted that there often is an intermediate optimum of allocation to reproductive meristems. The models assumed that reproductive allocation is constant over time. The second model, Eqs (2.5)–(2.7), showed that bang-bang reproduction has the highest fitness. Optimal control models of the resource budgets of annual plants often predict a similar allocation pattern as the latter model (Cohen, 1971; Perrin and Sibly, 1993). The great majority of plant species, however, do not follow either of the strategies: reproductive allocation is not concentrated in the end of the life span, although it is plastic.

What model assumptions produce this discrepancy? Real meristem systems have many characteristics that are not included in these models. There are several critical assumptions:

- It is assumed that a plant can supply any number of growth meristems with carbohydrates, mineral nutrients, and other resources for further growth. There are no resource limitations for exponential expansion of the meristem system. In reality, relative growth rate and meristem production decelerates after an early exponential phase (Geber, 1989).

- The seed production is assumed to be the same for each reproductive meristem. This is questionable, even as an approximation, especially for inflorescence meristems.
- It is assumed that there are no resource limitations for seed production irrespective of number of reproductive meristems. However, a large number of flowers may decrease seeds/flower ratio (Lee, 1988). Simultaneous mass flowering may also decrease pollination success per each individual flower and increase geitonogamy (Klinkhamer and de Jong, 1993). Simultaneous flowering and seed production has also other ecological consequences that may be beneficial or disadvantageous. Mass reproduction may attract seed predators or satiate them (Kelly and Sork, 2002). Dispersal ability may also be either increased or decreased due to mass blooming (Kelly and Sork, 2002).
- The role of meristems in the models is either to produce new meristems or seeds. Meristems have, however, many other functions, such as controlling plant ontogeny and architecture (Traas and Vernoux, 2002).
- Dormant buds have no selective role in the models above. However, the main function of producing nodes may often be to produce leaves and enhance height growth (Aarssen, 1995). Axillary meristems are then merely a byproduct of node production for other purposes and remain dormant. Dormant buds may also be important in preparing for environmental hazards, such as herbivory, frost, and drought damage (Tuomi et al., 1994; Aarssen, 1995).
- Optimal control models of resource allocation have shown that bang-bang reproduction is the most optimal strategy only under restricted conditions (Fox, 1992; Kozłowski, 1992). Gradual switches or several switches between vegetative and reproductive growth may be optimal if, for instance, the length of the growing season is unpredictable, if there are several unfavorable and favorable periods within season, if the cost of one offspring increases with the number of seeds produced simultaneously, if reproductive tissues are photosynthetic, or if storage capacity is limited. Analogous factors may affect the outcome of meristem models.

It is possible to release some of the assumptions in the models, whereas some other assumptions are directly incompatible with the idea that meristem allocation would determine reproductive allocation. For instance, if available resources are depleted by the demands imposed by currently active meristems, resource dynamics may completely dominate over meristem dynamics. Even in that case, it is possible that patterns of meristem allocation reflect underlying patterns of resource allocation, but it has no determining role in reproductive allocation. We will discuss the model assumptions in the next sections.

IX. The Impact of Meristem Allocation on Reproductive Allocation

Meristem allocation models suggest that plant allocation to different types of meristems may affect reproductive allocation. According to the models, early allocation to reproductive meristems instead of growth meristems may reduce the expansion of the shoot system and in this way later allocation to reproduction. The result is common for many allocation models, which predict a trade-off between early and late allocation to reproduction (Roff, 1992). In meristem allocation models, the availability of meristems limits reproduction. In other allocation models, carbon, mineral nutrients, or other resources may limit reproduction. However, the role of meristems as a limiting factor has some peculiarities. The limiting factor is the production of buds, small plant parts seemingly inexpensive to produce in terms of resource expenditure.

Several empirical studies have shown that meristem availability may affect reproductive allocation. Watson (1984) studied the meristem allocation of two experimental populations of water hyacinth *Eichhornia crassipes* (Pontederiaceae), an aquatic clonal herb. The flowering population had a much lower clonal growth rate and lower equilibrium density in a density-dependent population growth than the vegetative population. Resource limitation did not explain the result, but plants could supply new ramets with the necessary resources. Clonal expansion was meristem limited because flowering consumed apical meristems and decreased the production rate of new ramets.

Geber (1990) showed that meristem allocation affects reproductive allocation in *Polygonum arenastrum* (Polygonaceae), an annual herb. There was genotypic variation in the early allocation to flowering. Genotypes that had a high early allocation to flowering suffered from a cost in terms of low growth rate and fecundity later in life. The experimental plants did not produce any dormant meristems. All genotypes produced equal numbers of seeds per flowering metamer, which suggests that plants with high early reproductive allocation did not suffer from higher resource limitation than other plants. Plants were given ample water and fertilizer, which also suggests that it was unlikely that resource limitation would explain the result.

Some studies have not found any correlation among allocation to different types of meristems. Lehtilä *et al.* (1994) studied the bud dynamics of mountain birch *Betula pubescens* ssp. *czerepanovii* (Betulaceac) with population matrix models. They did not find any significant cost of flowering on production of buds. The rate of canopy expansion was similar for trees with and without inflorescences. There was, however, a 2–10% decrease of bud production rate in reproductive shoots, but statistical power was not sufficient to test for small costs. Morphological studies suggest that there may be meristematic costs of flowering in birches. Female inflorescences are located in short shoots. Short shoots are usually monopodial, and flowering terminates the shoot growth. However, short shoot growth may often continue from an axillary bud after flowering (Macdonald and Mothersill, 1983). Long shoots produce male inflorescences. Reproductive long shoot buds have fewer leaf primordia than vegetative long shoot buds (Macdonald *et al.,* 1984). Flowering should thus imply a meristem loss. Karlsson *et al.* (1996) found a significant cost of producing generative long shoots in mountain birch. Long shoots that had catkins produced fewer new long shoots from axillary buds, and the new long shoots had a reduced number of buds. Trees studied by Karlsson *et al.* (1996) had more inflorescences than those studied by Lehtilä *et al.* (1994), which may affect the expression of costs. Gross (1972) showed that intensive flowering may lead to a deterioration of vegetative growth in birches. These results indicate that reproductive allocation may affect later meristem allocation, but it is possible that underlying resource limitation may have been a causal factor.

Huber and During (2001) also did not find any cost of allocation to flowering meristems in a study of two *Trifolium* species. Although flowering from buds of the primary shoot was negatively correlated with branching from the primary shoot, in the whole-plant level the correlation between flowering and branching was not significant. Duffy *et al.* (1999) also did not observe any cost of flowering in terms of allocation to growth meristems. Instead, they observed a positive association between allocation to growth and reproductive meristems in *Arabidopsis thaliana.* In both studies, allocation to growth and flowering was negatively associated with allocation to inactive meristems.

X. Plasticity of Meristem Allocation

Plasticity of meristem allocation confers plants with the ability to avoid meristem limitation. If a plant can adaptively change the meristem allocation according to the variability of resources, it can take advantage of resource pulses. No plants have totally canalized meristem allocation, but it is possible that plasticity is constrained by plant allometry. If meristem allocation follows simple allometric rules, the proportion of each type of meristem will change according to a simple function through plant ontogeny. Regressions between meristem allocation and plant size, or regressions among the proportions of different types of meristems, show whether allometric rules determine meristem allocation (Bonser and Aarssen, 2001). Plants commonly have a threshold size for flowering which can be a part of the allometric function for meristem allocation.

Huber and Stuefer (1997) suggested that the effect of shading on branching of *Potentilla reptans* (Rosaceae), a stoloniferous herb, can be explained with allometry. Differences in branching patterns were similar for individuals at the same developmental stage, irrespective of shading treatment. Shading slowed down plant development, thus supporting the idea that shading affects branching predominantly through allometric effects.

Bonser and Aarssen (2001) studied the allometry of meristem allocation of *Arabidopsis thaliana*. Size differences were induced with nutrient treatments. The dominating allometric pattern was large size and was associated with increased allocation to reproductive and growth meristems and decreased allocation to inactive meristems. Allometry had a strong contribution to allocation patterns, but did not totally explain the effect of the environment. Allometry was strongest early in the development, i.e., plant size had a strong effect on meristem allocation. Allocation to flowering and branching was low in small young plants and high in large young plants. This may be interpreted as an adaptive response where growth and size are maximized early in development in high nutrient treatments.

Low plasticity of meristem allocation may be due to low responsiveness to environmental signals. Some traits may be fixed during a sensitive period early in the ontogeny and remain permanent for the rest of the lifetime. Novoplansky *et al.* (1994) showed that shading affected the branching pattern of the annual plant *Onobrychis squarrosa* (Fabaceae), so that a short period of shading (15–20 days) early in the growing season was sufficient to elicit the change in branching. The change was irreversible.

Many perennial plants have bud preformation, where many of the characteristics of the shoot system are determined one year or even several years before the actual growth and maturation of the shoots. Developmental preformation causes a delay in morphological responses to environmental variation. Diggle (1997) observed an extreme preformation of shoots of alpine *Polygonum viviparum*, where the shoot structure was preformed several years ahead. *Caltha leptosepala* (Ranunculaceae), another alpine herb, showed preformation where leaves were preformed two years and flowers one year before bud maturation (Aydelotte and Diggle, 1997). *Caltha leptosepala* expressed some plasticity through floral abortion and selective maturation of leaf primordia.

Preformation is common in cold climates (Aydelotte and Diggle, 1997). When the growing season is short, it seems to be difficult to respond to environmental variation through growth and reproduction in the same year. Instead, environmental variation affects storage and primordial structures of buds. Plants with preformation do not lack plasticity, but plastic reactions are stretched over several seasons.

Preformed structures do not always totally determine the shoot structure. In birch, shoot structure is preformed one year before bud break (Macdonald and Mothersill, 1983; Macdonald *et al.*, 1984). However, there are several stages after preformation period where the current environment may affect shoot growth. Early leaf damage may affect later shoot growth of the same season, in addition to the effects on the next season

(Ruohomäki *et al.*, 1997). Long shoot buds have preformed late leaves that develop throughout the summer. All leaf primordia do not always develop, similarly as in *Caltha leptosepala* (Aydelotte and Diggle, 1997). In addition, vegetative long shoots may produce "true" late leaves that are not preformed (Macdonald *et al.*, 1984). Furthermore, buds preformed to short shoots may develop to long shoots if neighboring long shoot buds are removed by winter browsing (Danell *et al.*, 1985). Despite preformation, the meristem system of birch is remarkably flexible. Removal of up to 35% of buds in winter moderately decreased the number of active shoots the next summer and changed the ratio of long shoots versus short shoots, but it did not affect the total leaf biomass (Lehtilä *et al.*, 2000).

Schmitt's (1993) study on *Impatiens capensis* (Balsaminaceae), an annual herb, demonstrated that there may be among-population variation in phenotypic plasticity of meristem allocation. One of the study populations came from an open habitat and the other one from a shady environment. Genotypes of both populations were grown in both high and low light. The time of anthesis was plastic for plants from the sunny population. In high light, flowering was delayed and meristems were allocated to branching in the early development of plants from the sunny population. In low light, flowering was early and branches fewer, resulting in a negative association between early reproductive allocation and later growth. Plants from the shady population started to flower earlier than plants from the sunny population, and the time of anthesis was not sensitive to light treatment. Light treatment did not affect branching of plants of the shady population. Donohue *et al.* (2000a,b, 2001) used transplantations at high and low density between populations, and again observed that individuals from the sunny population were more plastic. Donohue *et al.* (2000a, 2001) showed that selection favors plasticity in the sunny population, but not in the shady population, where high early mortality occurred. Irrespective of plant competition or shading, late-flowering plants had a low fitness in the shady population. Such pattern suggests that plasticity was adaptive in the sunny population.

XI. Major Genes of Meristem Allocation

The morphology of maize and its wild ancestor teosinte provides an example of evolution of meristem allocation patterns. In teosinte, axillary shoots are elongated terminating in a tassel (male inflorescence) and from these shoots ears (female inflorescences) develop. Domestication of teosinte resulted in *Zea mays* (Poaeceae) with a different meristem allocation pattern: the main axis terminates in a tassel and axillary shoots develop only from a few nodes and these terminate in ears. This represents repressed bud growth, supression of axillary branch formation, as well as meristem identity changes.

The difference in shoot architecture between teosinte and maize has been studied using a mutant of maize, *teosinte-branched1 (tb1)*, that closely resembles the teosinte architecture (Doebley *et al.*, 1997; Cubas *et al.*, 1999). A model supported by molecular and genetic data says that TB1 functions as a repressor of lateral growth that in maize would be expressed at a higher level in lateral shoot buds to suppress their outgrowth.

Another example of evolution of inflorescence architecture is *Jonopsidium acaule* (Brassicaceae). In *J. acaule*, which is closely related to inflorescence-bearing species, flowers develop from leaf axils in the rosette rather than from an inflorescence. Phylogenetic analysis suggests that *J. acaule* represents a rosette-flowering lineage that evolved from an inflorescence-bearing ancestor (Zunk *et al.*, 1996). Comparative studies of the expression of the *LFY* genes in Arabidopsis and *J. acaule* (Shu *et al.*, 2000) suggest that changes in the regulation of *LFY* expression could have played a role in the evolution of this rosette-flowering lineage.

The Arabidopsis gene *TERMINAL FLOWER1 (TFL1)* is yet another example of a gene that has a profound effect on meristem allocation (Bradley *et al.*, 1997). The expression of the gene prevents the development of floral identity of the shoot apical meristem. It thus allows for the elaboration of an inflorescence rather than the development of a terminal flower. In addition to this, it regulates the switch from vegetative to reproductive phase. Analyses have shown that *TFL1* delays the activation of *LFY* and another gene, *APETALA1*, both involved in inflorescence development.

These examples suggest that a small number of genes may control processes that have profound influence on shoot architecture and that the change in the expression of one of these genes would change the pattern of meristem allocation over evolutionary time.

XII. Resource Levels and Meristem Limitation

Geber (1990) suggested that meristem limitation is more common in rich environments than in poor environments. According to this idea, when resource levels are good, plants with few meristems cannot invest all available resources to growth and reproduction. Plants with many meristems can more effectively acquire and use resources. In poor environments, the number of meristems is not critical because fewer meristems are needed to use all available resources.

Donohue and Schmitt (1999) studied allocation patterns of *Impatiens capensis* in different competitive and light environments in a greenhouse. They observed that genetic trade-offs involving meristem allocation to branching were only expressed at low density. At low density, the genetic correlation between the number of branches and the number of flowers was negative. There was also a negative genetic correlation between the number of flowers and the number of dormant buds at both low and high density. Donohue *et al.* (2000b) carried out the competition experiment in two populations, one in a sunny site and the other in a shady site. As in Schmitt (1993), only genotypes of sunny population showed genetically based trade-offs, and they were only expressed at high density.

Donohue and Schmitt (1999) suggested that high density results in a large variation in resource acquisition among plants. Genotypes with high vigor were then better in producing both branches and flowers. Genotypes with lower vigor had a low competitive ability and produced fewer branches and flowers. However, at low density, there was less variation in resource intake. Since light and soil resources were plentiful, plant performance was determined more by the ability to use resources for growth and reproduction rather than to acquire more resources. Hence, as a consequence of high resource availability, the income gap among genotypes decreased.

It is often suggested that trade-offs are stronger when resource availability is low (Tuomi *et al.*, 1983; Reznick, 1985), but the findings of Donohue and Schmitt (1999) suggest that the situation is the opposite for meristematic costs. If high resource availability enables the activation of a large proportion of meristems, allocation to flowers and inflorescences may decrease the possibility of allocation to branching. If plants have many dormant meristems due to low resource availability, association between allocation to branches and to flowering may be weak or positive.

XIII. The Function of Dormant Buds

Many plants produce numerous dormant buds. The existence of dormant buds may be an indication that resource status limits growth. There are perhaps not enough resources to activate all meristems, so meristem availability cannot limit growth.

Dormant buds may well be an indication of resource limitation, but other possibilities also exist. Dormant buds may form a reserve that is mobilized if active shoots are lost. Several bud bank models have been presented to explain compensatory responses of plants to herbivores (van der Meijden, 1990; Vail, 1992; Tuomi *et al.*, 1994). The starting point has been to explain controversial observations of overcompensatory responses, according to which plants may have higher fitness when damaged than when they are left intact (Paige and Whitham, 1987; Lennartsson *et al.*, 1997). The general idea of the bud bank models is that when herbivory pressure is high and predictable, it is beneficial to produce dormant buds in locations protected from herbivores. Additional conditions for the evolution of overcompensation are that activation of dormant buds is not possible without a signal from damage and that probability of repeated damage is low (Nilsson *et al.*, 1996; Lehtilä, 2000).

The importance of dormant buds as a reserve depends on their ability to respond to environmental stimuli. It is common that older dormant buds senesce and cannot be activated even if resource levels are increased. In white clover, poor soil resources, defoliation, and plant competition affected axillary buds so much that sometimes only 5% of buds had a potential to become active (Newton and Hay, 1996). The fact that environmental conditions affect bud viability suggests that dormant buds have construction and maintenance costs.

Production of dormant buds may be a consequence of plant growth, without any direct adaptive value of dormant buds. Every leaf has one or several axillary buds. The pool of dormant buds may be a byproduct of leaf production when a plant increases the amount of photosynthetic machinery. The production of dormant buds may also be a byproduct of selection for apical dominance (Aarssen, 1995). Apical dominance affects height growth, which is important in plant competition, pollination, and seed dispersal. Height growth may happen by internode elongation or by increasing the number of nodes. The latter increases the number of dormant buds.

More generally, meristem allocation may affect plant fitness through other processes than directly via reproductive allocation. The structure of plant canopy is determined by meristem numbers, meristem fates, branching angles, and internode lengths. Canopy structure has several characteristics that have adaptive value. It affects competitive ability for light, pollination and seed dispersal, structural stability, transport of water and assimilates, and light interception (Aarssen, 1995). These functions are affected by the number and placement of different types of meristems, and the optimal solution varies according to the relative importance of different functions.

XIV. Meristem Allocation as a Surrogate in Estimation of Resource Allocation

When biomass, carbon, energy, or nutrients are used as a currency of allocation, it is difficult to make a distinction between reproductive and vegetative parts of a plant. Additional problems are a dynamic resource use and resources with quick turnover. Meristems, on the other hand, are distinctively reproductive, inactive, or growing. Contrary to many other methods, meristem counting is nondestructive, which is important when allocation patterns are followed over time. Remembering the fact that the availability of reproductive meristems can actually constrain reproductive output, meristem allocation may be a good option in measuring reproductive allocation.

On the other hand, meristem counting can be very laborious for large plants. Meristems may be inconspicuous, especially in older parts of perennial plants where buds may be hidden by secondary growth. Although such old buds can often be ignored because they have a low probability of activation, large canopies demand efficient sampling strategies to estimate meristem number and stage structure.

If meristem counting is used as a surrogate for measuring resource allocation, meristem stages and fates must reflect allocation of other resources. The relationship between resource allocation and meristem allocation is simple if resource investment per meristem is constant, at least among meristems of the same type. If, on the other hand, a plant invests a varying amount of resources in meristems that belong to the same type, it is difficult to assess resource allocation patterns from meristem allocation patterns. The latter is probably true for most plants. The variation is largest in inflorescence meristems. Resource allocation to a single inflorescence varies depending on how many flowers there are in the inflorescence, and the investment is plastic according to resources available. Counting flowers in addition to inflorescences may help, but even when the number of flowers is known, the likelihood of flower or fruit abortion, the number of seeds and the investment for each seed can vary considerably among reproductive meristems.

Whether these factors substantially confound the relationship between meristem allocation and resource allocation is species dependent. At best, meristem allocation gives us an insight of allocation to reproduction for an easy comparison among genotypes, experimental groups or species, although it does not give the same quantitative power to study the ecophysiological background as measurements of reproductive investments with energy, carbon, or mineral nutrients as a currency.

XV. Conclusions

The fact that meristem availability limits reproduction, at least in some conditions, has clearly demonstrated that architectural and structural aspects cannot be ignored in studies of plant resource allocation. The principle of resource allocation states that evolutionary trajectories of reproductive effort are constrained by resource availability. If meristem allocation plays a role in reproductive allocation, there is a further constraint – if there are ample resources, flower and seed production may be limited by the ability to produce reproductive meristems. However, since meristem limitation seems to depend on environmental conditions and species characteristics, the measurement of meristem allocation is usually not a shortcut in assessing reproductive allocation. Many plant species clearly produce enough meristems and live under a resource-limited situation, hence they presumably seldom meet meristem limitation. Other plants can cope with meristem limitation by plasticity of meristem allocation. Plants have both short-term developmental and long-term evolutionary potential to avoid meristem limitation.

Further studies are needed to elucidate possible costs of plasticity and constraints in the evolution of meristem allocation. Meristem counts may work as a surrogate for other measures of resource allocation, but this depends on the correspondence of meristem allocation with the allocation of other resources, which is not clear, and on logistics of particular experiments – does it take a long time to count meristems? More studies are needed to clarify the mechanistic basis of meristem allocation patterns. Advancements in developmental biology allowed by new molecular methods will provide some of the missing foundations needed in this area.

Acknowledgments

The work was partially financed by Swedish Research Council and Östersjöstiftelsen.

References

Aarssen L. W. (1995) Hypotheses for the evolution of apical dominance in plants: implications for the interpretation of overcompensation. *Oikos* **74**: 149–156.

Aydelotte A. R. and Diggle P. K. (1997) Analysis of developmental preformation in the alpine herb *Caltha leptosepala* (Ranunculaceae). *Amer J Bot* **84**: 1646–1657.

Bazzaz F. A., Carlson R. W. and Harper J. L. (1979) Contribution to reproductive effort by photosynthesis of flowers and fruits. *Nature* **279**: 554–555.

Blázquez M. A. (2000) Flower development pathways. *J Cell Science* **113**: 3547–3548.

Bonser S. P. and Aarssen L. W. (1996) Meristem allocation: a new classification theory for adaptive strategies in herbaceous plants. *Oikos* **77**: 347–352.

Bonser S. P. and Aarssen L. W. (2001) Allometry and plasticity of meristem allocation throughout development in *Arabidopsis thaliana*. *J Ecol* **89**: 72–79.

Bradley D., Ratcliffe O., Vincent C., Carpenter R. and Coen E. (1997) Inflorescence commitment and architecture in *Arabidopsis*. *Science* **275**: 80–83.

Brand U., Hobe M. and Simon R. (2001) Functional domains in plant shoot meristems. *BioEssays* **23**: 134–141.

Cohen D. (1971) Maximizing final yield when growth is limited by time or by limiting resources. *J Theor Biol* **33**: 299–307.

Cubas P., Lauter N., Doebley J. and Coen E. (1999) The TCP domain: a motif found in proteins regulating plant growth and development. *Plant J* **18**: 215–222.

Danell K., Huss-Danell K. and Bergström R. (1985) Interactions between browsing moose and two species of birch in Sweden. *Ecology* **66**: 1867–1878.

Diggle P. K. (1997) Extreme preformation in alpine *Polygonum viviparum*: an architectural and developmental analysis. *Amer J Bot* **84**: 154–169.

Doebley J., Stec A. and Hubbard L. (1997) The evolution of apical dominance in maize. *Nature* **386**: 485–488.

Donohue K., Hammond Pyle E., Messiqua D., Heschel M. S. and Schmitt J. (2000a) Density dependence and population differentiation of genetic architecture in *Impatiens capensis* in natural environments. *Evolution* **54**: 1969–1981.

Donohue K., Messiqua D., Hammond Pyle E., Heschel M. S. and Schmitt J. (2000b) Evidence of adaptive divergence in plasticity: density- and site-dependent selection on shade avoidance responses in *Impatiens capensis*. *Evolution* **54**: 1956–1968.

Donohue K., Hammond Pyle E., Messiqua D., Heschel M. S. and Schmitt J. (2001) Adaptive divergence in plasticity in natural populations of *Impatiens capensis* and its consequences for performance in novel habitats. *Evolution* **55**: 692–702.

Donohue K. and Schmitt J. (1999) The genetic architecture of plasticity to density in *Impatiens capensis Evolution* **53**: 1377–1386.

Duffy N. M., Bonser S. P. and Aarssen L. W. (1999) Patterns of variation in meristem allocation across genotypes and species in monocarpic Brassicaceae. *Oikos* **84**: 284–292.

Emery R. J. N., Longnecker N. E. and Atkins C. A. (1998) Branch development in *Lupinus angustifolia* L. II. Relationship with endogenous ABA, IAA and cytokinins in axillary and main stem buds. *J Exp Bot* **49**: 555–562.

Fletcher J. C. (2002) Coordination of cell proliferation and cell fate decisions in the angiosperm shoot apical meristem. *BioEssays* **24**: 27–37.

Fox G. A. (1992) Annual plant life histories and the paradigm of resource allocation. *Evol Ecol* **6**: 482–499.

Gadgil M. and Bossert W. H. (1970) Life historical consequences of natural selection. *Am Nat* **104**: 1–24.

Garrison R. (1949a) Origin and development of axillary buds: *Betula papyrifera* March. and *Euptelea polyandra* Sieb. et Zucc. *Amer J Bot* **36**: 205–213.

Garrison R. (1949b) Origin and development of axillary buds: Syringa vulgaris L. *Amer J Bot* **36**: 257–266.

Garrisson R. (1955) Studies in the development of axillary buds. *Amer J Bot* **42**: 257–266.

Geber M. A. (1989) Interplay of morphology and development on size inequality: a *Polygonum* greenhouse study. *Ecol Monogr* **59**: 267–288.

Geber M. A. (1990) The cost of meristem limitation in *Polygonum arenastrum*: negative genetic correlations between fecundity and growth. *Evolution* **44**: 799–819.

Grbic V. and Bleeker A. B. (2000) Axillary meristem development in *Arabidopsis thaliana*. *Plant J* **21**: 215–223.

Grime J. P. (1979) *Plant Strategies and Vegetation Processes*. Wiley, New York.

Gross H. L. (1972) Crown deterioration and reduced growth associated with excessive seed production by birch. *Can J Bot* **50**: 2431–2437.

Hempel F. D., Zambryski P. C. and Feldman L. J. (1998) Photoinduction of flower identity in vegetatively biased primordia. *Plant Cell* **10**: 1663–1675.

Huber H. and During H. J. (2001) No long-term costs of meristem allocation to flowering in stoloniferous *Trifolium* species. *Evol Ecol* **14**: 731–748.

Huber H. and Stuefer J. F. (1997) Shade-induced changes in the branching pattern of a stoloniferous herb: functional response or allometric effect? *Oecologia* **110**: 478–486.

Jones M. (1985) Modular demography and form in silver birch. In *Studies on Plant Demography: A Festschrift for John L. Harper* (J. White, ed.) pp. 223–237, Academic Press Inc, London.

Karlsson P. S., Olsson L. and Hellström K. (1996) Trade-offs among investments in different long-shoot functions – variation among mountain birch individuals. *J Ecol* **84**: 915–921.

Kelly D. and Sork V. L. (2002) Mast seeding in perennial plants: why, how, where? *Ann Rev Ecol Syst* **33**: 427–447.

Kerstetter R. A. and Hake S. (1997) Shoot meristem formation in vegetative development. *Plant Cell* **9**: 1001–1010.

Klinkhamer P. G. L. and de Jong T. J. (1993) Attractiveness to pollinators: a plant's dilemma. *Oikos* **66**: 180–184.

Kozłowski J. (1992) Optimal allocation of resources to growth and reproduction: implications for age and size at maturity. *Trends Ecol Evol* **7**: 15–19.

Lee T. D. (1988) Patterns of fruit and seed production. In *Plant Reproductive Ecology: Patterns and Strategies* (J. Lovett Doust and L. Lovett Doust, eds.) pp. 179–202, Oxford University Press, New York.

Lehtilä K. (2000) Modelling compensatory regrowth with bud dormancy and gradual activation of buds. *Evol Ecol* **14**: 315–330.

Lehtilä K., Tuomi J. and Sulkinoja M. (1994) Bud demography of the mountain birch *Betula pubescens* ssp. *tortuosa* near treeline. *Ecology* **75**: 945–955.

Lehtilä K., Haukioja E., Kaitaniemi P. and Laine K. A. (2000) Allocation of resources within mountain birch canopy after simulated winter browsing. *Oikos* **90**: 160–170.

Lennartsson T., Tuomi J. and Nilsson P. (1997) Evidence for an evolutionary history of overcompensation in the grassland biennial *Gentianella campestris* (Gentianaceae). *Am Nat* **149**: 1147–1155.

Long J. and Barton K. (2000) Initiation of axillary and floral meristems in *Arabidopsis*. *Dev Biol* **218**: 341–353.

MacArthur R. H. and Wilson E. O. (1967) *The Theory of Island Biogeography.* Princeton University Press, Princeton, New Jersey.

Macdonald A. D. and Mothersill D. H. (1983) Shoot development in *Betula papyrifera.* I. Shortshoot organogenesis. *Can J Bot* **61**: 3049–3065.

Macdonald A. D., Mothersill D. H. and Caesar J. C. (1984) Shoot development in *Betula papyrifera.* III. Long-shoot organogenesis. *Can J Bot* **62**: 437–445.

Maillette L. (1982) Structural dynamics of silver birch. II. A matrix model of the bud population. *J Appl Ecol* **19**: 219–238.

Majumdar G. P. (1942) The organisation of the shoot in *Heracleum* in the light of development. *Ann Bot* **6**: 49–83.

Meijden E. van der (1990) Herbivory as a trigger for growth. *Funct Ecol* **4**: 597–598.

Miguel L. C., Longnecker N. E., Ma Q., Osborne L. and Atkins C. A. (1998) Branch development in *Lupinus angustifolia* L. I. Not all branches have the same potential growth rate. *J Exp Biol* **49**: 547–553.

Morelli G. and Ruberti I. (2000) Shade avoidance responses. Driving auxin along lateral routes. *Plant Physiol* **122**: 621–626.

Napoli C. A., Beveridge A. B. and Snowden K. C. (1999) Reevaluating concepts of apical dominance and the control of axillary bud outgrowth. *Curr Topics Dev Biol* **44**: 127–169.

Newton P. C. D. and Hay M. J. M. (1996) Clonal growth of white clover: factors influencing the viability of axillary buds and the outgrowth of a viable bud to form a branch. *Ann Bot* **78**: 111–115.

Nilsson P., Tuomi J. and Åström M. (1996) Even repeated grazing may select for overcompensation. *Ecology* **77**: 1942–1946.

Novoplansky A., Cohen D. and Sachs T. (1994) Responses of an annual plant to temporal changes in light environment: an interplay between plasticity and determination. *Oikos* **69**: 437–446.

Ohto M., Onai K., Furukawa Y., Aoki E., Araki T. and Nakamura K. (2001) Effects of sugar on vegetative developement and floral transition in Arabidopsis. *Plant Physiol* **127**: 252–261.

Paige K. N. and Whitham T. G. (1987) Overcompensation in response to mammalian herbivory: the advantage of being eaten. *Am Nat* **129**: 407–416.

Perrin N. and Sibly R. M. (1993) Dynamic models of energy allocation and investment. *Ann Rev Ecol Syst* **24**: 379–410.

Reekie E. G. (1999) Resource allocation, trade-offs, and reproductive effort in plants. In *Life History Evolution in Plants* (T. O. Vuorisalo and P. K. Mutikainen eds.) pp. 173–193, Kluwer Academic Publishers, Dordrecht.

Remphrey W. R. and Steeves T. A. (1984) Shoot ontogeny in *Arctostaphylos uva-ursi* (bearberry): origin and early development of lateral vegetative and floral buds. *Can J Bot.* **62**: 1933–1939.

Reznick D. (1985) Costs of reproduction: an evaluation of the empirical evidence. *Oikos* **44**: 257–267.

Roff D. A. (1992) *The Evolution of Life Histories.* Chapman and Hall, New York.

Ruohomäki K., Haukioja E., Repka S. and Lehtilä K. (1997) Leaf value: effects of damage to individual leaves on growth and reproduction of mountain birch shoots. *Ecology* **78**: 2105–2117.

Schmitt J. (1993) Reaction norms of morphological and life-history traits to light availability in *Impatiens capensis. Evolution* **47**: 1654–1668.

Sharman B. C. (1942) Developmental anatomy of the shoot of *Zea mays* L. *Ann Bot* **6**: 245–282.

Shu G., Amaral W., Hileman L. C. and Baum D. A. (2000) LEAFY and the evolution of rosette flowering in violet cress (*Jonopsidium acaule,* Brassicaceae) *Am J Bot* **87**: 634–641.

Simpson G. G., Gendall A. R. and Dean C. (1999) When to switch to flowering. *Ann Rev Cell Dev Biol* **99**: 519–550.

Smith H. (1995) Physiological and ecological function within the phytochrome family. *Ann Rev Plant Physiol Plant Mol Biol* **46**: 289–315.

Smith H. and Whitelam G. C. (1997) The shade avoidance syndrome: multiple response mediatied by multiple phytochromes. *Cell Env* **20**: 840–844.

Tamas I. A. (1995) Hormonal regulation of apical dominance. In *Plant Hormones* (P. J. Davies, ed.) pp. 572–597, Kluwer Academic Publishers, Dordrecht.

Tantikanjana T., Yong J. W. H., Letham D. S., Griffith M., Hussain M., Ljung K., Sandberg G. and Sundaresan V. (2001) Control of axillary bud initiation and shoot architecture in *Arabidopsis* through the SUPERSHOOT gene. *Genes Dev* **15**: 1577–1588.

Traas J. and Vernoux T. (2002) The shoot apical meristem: the dynamics of a stable structure. *Phil Trans R Soc Lond B* **357**: 737–747.

Tuomi J., Hakala T. and Haukioja E. (1983) Alternative concepts of reproductive effort, cost of reproduction, and selection in life-history evolution. *Amer Zool* **23**: 25–34.

Tuomi J., Nilsson P. and Åström M. (1994) Plant compensatory responses: bud dormancy as an adaptation to herbivory. *Ecology* **75**: 1429–1436.

Vail S. G. (1992) Selection for overcompensatory plant responses to herbivory: a mechanism for the evolution of plant-herbivore mutualism. *Am Nat* **139**: 1–8.

Wardlaw C. W. (1943) Experimental and analytical studies of Pteridophytes. I. Preliminary observations on the development of buds on the rhizome of the ostrich fern (*Matteuccia struthiopteris* Tod.). *Ann Bot* **26**: 171–187.

Watson M. A. (1984) Developmental constraints: effects on population growth and patterns of resource allocation in a clonal plant. *Am Nat* **123**: 411–426.

Williams G. C. (1966) Natural selection, the costs of reproduction, and a refinement of Lack's principle. *Am Nat* **100**: 687–690.

Zunk K., Mummenhoff M., Koch M. and Hurka H. (1996) Phylogenetic relationships of *Thlaspi* s.l. (subtribes Thlaspidiane, Lepideae) and allied genera based on chloroplast DNA restriction-site variation. *Theor Appl Genet* **92**: 375–381.

3

It Never Rains but then it Pours: The Diverse Effects of Water on Flower Integrity and Function

Candace Galen

I. Introduction

The summer of 2002 was a stark reminder of the toll in ecological and environmental devastation taken by climate extremes. While torrential rains caused the worst floods that northern Europe had experienced in at least a century, a lack of precipitation in large parts of North America resulted in the most severe drought in the past 107 years. For crop cultivars and native plant species, the timing and magnitude of precipitation are critical in mediating its impact on seed and fruit production (Passioura, 1996). When drought or heavy rainfall coincide with anthesis and pollination, catastrophic losses of grain and fruit crops often ensue (e.g., Chiarello and Gulmon, 1991; Westgate and Peterson, 1993; Hashem *et al.*, 1998; Nam *et al.*, 2001). Drought is a major cause of losses to historic and modern crop production (Ceccarelli and Grando, 1996; Cushman, 2001). In nature too, detrimental effects of drought propagate through food webs, threatening the viability of bird populations that depend on floral or seed resources (Grant, 1985; Stiles, 1992). These examples underscore the importance of understanding the water relations of flowering.

In this chapter, I examine how water availability affects the diversity of flower functions integral to the process of sexual reproduction (Fig. 3.1). I then review the evidence that water use by flowers is costly to vegetative growth. Third, I consider in detail how water acts to moderate other sources of environmental stress during anthesis (Fig. 3.1). The second half of the chapter moves from a functional perspective to an evolutionary one. I address three major ways in which water relations might influence floral evolution. First, I examine how selection to reduce the water cost of flowers

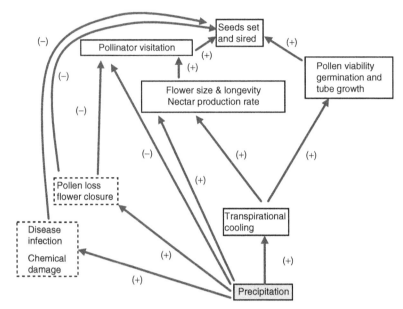

Figure 3.1 Multiple causal pathways linking water input during flowering to plant reproductive success. Red lines and dashed boxes indicate negative effects of water mediated through environmental conditions during flowering. Blue lines and solid boxes indicate positive effects of water as a limiting resource for flower functions. (See color plate.)

may lead to reductions in components of floral display (resource cost hypothesis, Galen, 1999). Next, I consider the adaptive significance of floral traits that influence the flower's water relations. I propose that floral traits may be targets of selection through parental environmental effects if they moderate the amount of water in the floral micro-environment or optimize the use of that water for temperature control. Last, I suggest that the plastic responses of floral traits to moisture regimes may alter the ecological interactions between plants and their animal pollinators. I show that water stress can affect the mode of pollinator-mediated selection on floral traits indirectly by altering floral rewards and/or flowering phenology.

II. The Functional Ecology of Water in the Life of a Flower

A. Water Use by Flowers

Water is used for multiple functions throughout the flower's life: cell division and cell expansion in the perianth depend on turgor maintenance, which in turn is driven by water influx into the cell vacuole (Yamane, Kawabata and Sakiyama, 1991). These cellular processes require an influx of water from the vegetative body of the plant during bud expansion and flower opening (Minter and Lord, 1983; Mohan Ram and Rao, 1984; Galen *et al.*, 1999). Because phloem rather than xylem supplies water to flowers, energetic costs associated with the movement of water into and through flowers are likely to exceed similar costs for leaves (De la Berra and Nobel, 2004). When water flow into floral meristems is curtailed under drought, rates of bud abortion for wild and domesticated plant species increase dramatically (e.g., Waller, 1980; Minter and Lord, 1983; Boot *et al.*, 1986; Smith-Huerta and Vasek, 1987; Mouhouche *et al.*, 1998; Andersson, 2000; Kokubun *et al.*, 2001). Drought stress induces ethylene synthesis, causing flowers to abscise prematurely. In *Capsicum annuum*, ethylene evolution in flowers increases 40-fold under drought stress (Jaafar *et al.*, 1998). Flowers that open despite drought often exhibit premature cell death and senescence (e.g., Westgate and Peterson, 1993; Panavas *et al.*, 1998; Arathi *et al.*, 2002). The signaling pathway from water deficit to bud or flower abortion also appears to involve the induction of abscisic acid (ABA) synthesis in response to drought. For example, in cotton, water deficit causes a 66% increase in the ABA content of flowers (Guinn *et al.*, 1990). Although some studies suggest that ABA in flowers acts to reduce membrane permeability and cause premature organ senescence (LePage-Degivry *et al.*, 1991; Asmussen, 1993; Minter and Lord, 1983; Panavas *et al.*, 1998), this conclusion may be premature. Recent work shows that ABA induction under drought in vegetative shoots actually maintains shoot growth in drying soils by limiting ethylene production (reviewed in Sharp and LeNoble, 2002). However, since flowers and fruit are isolated hydraulically from the transpiration stream of the xylem, inferences about flower behavior under drought based on leaf responses may be unwise (Davies *et al.*, 2000).

In addition to its role in turgor maintenance within floral organs, water has special molecular properties that make it extremely valuable to flower integrity under temperature extremes (Fig. 3.1). Because of its heat absorbing capacity, water is used for temperature regulation in flowers as well as leaves. In plants of hot tropical regions, transpirational cooling associated with water loss from the corolla enhances ovule survival by reducing pistil temperature (Patiño and Grace, 2002). Water also plays a key role in plant pollination – the quantity and sugar concentration of nectar for animal pollinators are highly sensitive to precipitation and humidity regimes (Fig. 3.1, Bertsche, 1983; Zimmerman, 1983; Zimmerman and Pyke, 1988; Boose, 1997; Carroll *et al.*, 2001). In arid environments, insect pollinators collect nectar for both its water content and energy value (Willmer, 1986, 1988). Water gradients serve as a guidance system in the style for pollen tube growth: disruption of such mechanisms under drought or flooding may reduce ovule fertilization rates despite plentiful compatible pollen receipt (Zuberi and Dickinson, 1985; Lush *et al.*, 1998). Last, for plant species of aquatic or wet terrestrial habitats, where animal pollinators and wind are unreliable, water is thought to serve as a mode of pollen transport (hydrophily, Faegri and van der Pijl, 1971; Catling, 1980; Sheviak, 2001).

B. The Water Cost of Flowers

Unlike carbon or nutrients, water cannot be directly taken up or resorbed from senescing organs, potentially magnifying its allocation cost. The water supply for developing and mature flowers comes at the direct expense of concurrent vegetative growth and maintenance. Indeed, the impact of drought on vegetative growth is often greatest when it occurs at the time of flowering (Chiarello and Gulmon, 1991; Passioura, 1996; Nam *et al.,* 2001).

That flowers are costly to the water budget of the plant is supported by geographical patterns of sex allocation in relation to drought stress. The frequency of self-pollination coupled with reduced allocation to floral display and flower longevity correlates with habitat aridity (Guerrant, 1984; but see Jonas and Geber, 1999). In subdioecious species made up of male, female, or hermaphroditic sex morphs, unisexuality typically increases with habitat aridity (Barrett, 1992; Costich, 1995; Weller *et al.,* 1995; Case and Barrett, 2001). This trend suggests that the water cost of simultaneous allocation to both male and female sexual functions imposes a fitness disadvantage for plants with hermaphroditic flowers that favors the maintenance of dioecy under drought (Webb, 1999). In subdioecious *Wurmbea dioica,* cosexual plants have reduced rates of leaf transpiration and higher water use efficiency compared to unisexual individuals, suggesting a history of selection for conservative water use under the added water demand of maintaining larger hermaphroditic flowers (Case and Barrett, 2001). Exacerbated allocation costs of flowering are predicted in arid environments, especially for plants that invest heavily in pollinator attraction and nectar reward (Nobel, 1977; Whiley *et al.,* 1988). In the alpine wildflower, *Polemonium viscosum,* production of large showy flowers reduces leaf water status and photosynthetic rate for plants flowering under dry conditions, but not for plants flowering under more mesic conditions (Galen *et al.,* 1999).

Nonetheless, detecting an increase in the physiological cost of flowering due to drought stress may be difficult for at least three reasons. First, the direct effects of water limitation on leaf function and vegetative growth under extreme drought may overwhelm additional indirect costs of resource allocation to flowering. Second, flowers and fruits represent carbon sinks that may enhance the rate of leaf photosynthesis, potentially masking or compensating for allocation costs (reviewed in Laporte and Delph, 1996). Third, in many plant species, flower number, size, and nectar production exhibit phenotypically plastic reductions under drought that alleviate allocation costs to vegetative growth (Minter and Lord, 1983; Smith-Huerta and Vasek, 1987; Boose, 1997; Andersson, 2000; Carroll *et al.,* 2001; Elle and Hare, 2002). Novel experimental approaches in which hormonal manipulations (e.g., GA application, Lord, 1980) are used to disrupt plastic responses to drought stress may reveal both the allocation costs associated with water use during flowering and the adaptive significance of phenotypically plastic reductions in flower size and/or number.

C. Water as a Regulator of Flower Microclimate

Water not only serves as a limiting resource for flower initiation, growth, and maintenance, but moderates critical aspects of the physical and chemical environment in which sexual reproduction takes place. Many effects of water on flower integrity involve its use in the maintenance of an optimal intrafloral microclimate. Atmospheric water in the form of water vapor or rainfall can have a profound influence on the quality of the flower microclimate during pollination (Fig. 3.1, Corbet, 1990). Low relative humidity during pollen maturation has negative effects on pollen germinability (Gilissen, 1977; Turner, 1993). Such negative paternal environmental effects could exert selection on traits like flowering phenology and anther position that affect the microclimate surrounding the stamens. Rainfall can have additional positive and negative effects on flower integrity depending on its timing and magnitude. For example, severe downpours have negative effects on male fertility and female function by washing pollen out of flowers and causing the remaining grains to burst prematurely (Eisokowitch and Woodell, 1974; Lisci *et al.,* 1994; Bynum and Smith, 2001; Huang *et al.,* 2002). Conversely, light rain or mist in the evening can buffer flower organs from overnight frost damage in the early spring. Liquid water provides this protection by releasing heat as the temperature drops, a phenomenon exploited to avert frost damage in flower and fruit crop cultivation. The agricultural practice of using overhead sprinklers to delay flowering in orchard crops is based on the capacity of water vapor to moderate bud temperature – evaporative cooling by buds delays flowering by 2–3 weeks in various fruit crops (Hewett and Young, 1980). In nature, cooling through evapotranspiration from mature flower surfaces lowers pistil and anther temperatures reducing the risk of thermal damage to ovules and pollen under hot conditions (Patiño and Grace, 2002). It is worth emphasizing that these impacts of water on flower function and integrity are unrelated to soil moisture. In other words, they represent direct effects of water on microclimate quality rather than on resource availability during flower growth and maintenance (Fig. 3.1). Studies that focus solely on water (or drought) as a resource factor during plant reproduction may consequently overlook significant environmental effects of water on male and female function.

D. Water as a Conduit for Environmental Sources of Flower Damage

As rainfall promotes the deposition of airborne fungal and bacterial spores, the timing of precipitation can determine the susceptibility of flowers to disease infection (Dodd *et al.,* 1991; Hildebrand *et al.,* 2001). Crab apple flowers exposed to the pathogenic bacteria *Erwinia amylovora* become diseased only when kept wet during inoculation (Figure 3.1, Pusey, 2000). Flowers readily intercept and absorb dissolved atmospheric pollutants delivered in rainfall as well with potentially devastating effects for plant reproduction (acid rain, Monson *et al.,* 1992). Stigma receptivity, pollen germination, and pollen tube growth are all pH sensitive processes that are disrupted by acid rain (Cox, 1984). Floral traits can modify the susceptibility of flowers and reproductive organs to damage from dissolved pollutants. In a study of morphological factors affecting plant exposure to acid rain, Monson *et al.* (1992) showed that flowers have high wettability defined as the mass of water retained per unit area upon immersion after all dripping has ceased. They report that inflorescences comprised of numerous large flowers make a major contribution to plant water retention per unit biomass under simulated rainfall. It follows that large floral displays adapted for pollinator attractiveness may inadvertently make plants more vulnerable to dissolved pollutants (an evolutionary trap, Schlaepfer, Runge, and Sherman, 2002). As with environmental effects of water on flower function, conduit effects represent a source of stress to reproductive organs that is unrelated to the availability of water as a resource, but follows from the biochemical (solvent) properties of liquid water.

III. Water Relations and the Evolution of Floral Traits

A. Floral Traits as Resource Sinks: The Resource Cost Hypothesis

Under arid conditions, allocation of water to support floral display can be costly to future growth and survival. Nobel (1977) modeled costs associated with diversion of water from leaves to inflorescences of the desert succulent, *Agave deserti.* His analysis suggests that the water cost of flowering is sufficient to explain the death of the vegetative plant after a single bout of reproduction. The resource cost hypothesis (Galen, 1999) postulates that when allocation costs of flower production and maintenance have significant fitness consequences, they may select for reductions in floral display countering selection exerted by animal pollinators. Ashman and Schoen (1994) make a similar argument for the impact of flower maintenance costs on the evolution of floral longevity. While their hypothesis addresses the physiological costs associated with the amount of time spent in pollen transfer rather than pollinator attraction per se, water figures prominently in both trade-offs. Water use by flowers may impose two simultaneous resource costs on vegetative organs: direct competition for water as a resource and exacerbated demand for photosynthate to support the active transport of water into flowers through the phloem (De la Berra and Nobel, 2004).

In the alpine wildflower, *Polemonium viscosum,* plants with large showy flowers obtain higher pollination success than plants with smaller less conspicuous flowers, but exhibit more negative leaf water potentials and reduced photosynthetic rates under dry conditions (Galen, 1989; Galen *et al.*, 1999). The combined production of large flowers and heavy seed crops incurs a survival cost for plants flowering under drought stress, but not for plants provided with supplemental water (Galen, 2000). Smaller flowered plants incur no such trade-off under drought. These results support the resource cost hypothesis and suggest that small flower size may be maintained by viability selection in drought-prone habitats.

Variation in flower size among populations of *P. viscosum* is also consistent with the resource cost hypothesis. I measured the flower of randomly chosen plants in five geographically isolated populations of *P. viscosum* in Colorado and Arizona during June, 1985 (range of 16–40 plants per population; for further details, see Galen *et al.*, 1987). Correlation analysis revealed a strong relationship between population mean flower length and the 50-year average for total June precipitation (Western Regional Climate Center, 2002) at the closest weather station (Spearman's $r = 0.90$, $P < 0.0374$; Fig. 3.2, Galen, unpublished data). This geographical pattern further

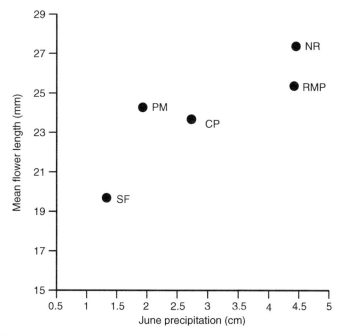

Figure 3.2   Relationship between water availability at the time of flowering as indexed by the historical average (1948–2001) for total June precipitation (cm), and average flower length (mm) in five geographically isolated populations of the alpine skypilot, *Polemonium viscosum* in the South-Central Rocky Mountains, USA. Site abbreviations are: (SF, San Francisco Peaks, AZ; PM, Pennsylvania Mountain, CO; CP, Cumberland Pass, CO; NR, Niwot Ridge, CO; RMP, Rocky Mountain National Park, CO). The relationship between water availability and flower size is significant (Spearman's nonparametric $r = 0.90$, $P < 0.0374$, Galen, unpublished data).

supports the role of drought in selecting for small flowers in *P. viscosum*. While the relationship between flower length and rainfall is based on phenotypic variation, it is likely to reflect genetic differentiation. Flower length in at least one population of *P. viscosum* has high heritability and is genetically correlated with other flower dimensions (Galen, 1996).

Comparative studies using phylogenetically independent contrasts to test for evolutionary divergence in floral traits between sister taxa of arid versus mesic habitats, provide additional insights into the adaptive significance of drought stress in the evolution of floral display. For example, Weller *et al.* (1995) mapped habitat aridity and breeding system onto a phylogeny of the endemic Hawaiian lineage containing *Schiedea* and *Alsindendron* (Caryophyllaceae). Their analysis revealed an association between dry habitats and the evolution of small unisexual flowers, supporting the role of drought in favoring reduced allocation to floral display. It would be especially interesting to map aspects of plant physiology that are involved in drought tolerance (e.g., water use efficiency, cellular osmotic potential, maximum stomatal conductance) onto phylogenies of this kind. Such studies could illuminate the role of correlated selection on plant functional traits and floral traits in driving local adaptation to dry habitats.

B. Floral Traits and Water in the Flower Microclimate: Parental Environmental Effects

An astonishing number of floral traits that have traditionally been considered solely to represent adaptations for pollinator attractiveness have an impact on flower microclimate and, through it, on plant water relations. Flower height, orientation, and movement, along with petal closure, petal surface area and corolla shape influence the exposure of pistils and stamens to water with implications for paternal and maternal components of reproductive success (Levin and Watkins, 1984; Sklenar, 1999; Bynum and Smith, 2001; Huang *et al.,* 2002; Patiño and Grace, 2002; Armstrong, 2002; Galen and Stanton, 2003). Some traits act directly to limit water damage to sexual organs. For example, the trichome fringe on petals of floating *Nymphoides* flowers creates an air bubble that protects the sexual organs from submersion (Armstrong, 2002). The bubble generates an upward force that counters pollinator body weight, maintaining the position of the flower on the water surface during pollinator visits. Other traits allow flowers to use water in maintaining optimal temperatures for pollen germination and tube growth. The parasol shape of tropical convolvulaceous flowers turns the corolla into a radiation shield that reduces, but does not eliminate water loss. Along with shade from the corolla, evapotranspiration through stomata in the corolla tube contributes to evaporative cooling of the flower microclimate, protecting male and female sex organs from damage under the hot humid conditions of the plant's natural habitat

(Patiño and Grace, 2002). When grease is applied to the corolla of *Ipomoea pes-capre* to prevent evapotranspiration, temperature of the gynoecium rises by about 5°C, exceeding a temperature threshold for loss of viability in sepals, female sexual organs, and pollen grains after only an hour.

Heliotropism or solar tracking also enhances the rate of flower evapotranspiration with implications for plant water relations and intrafloral temperature regulation. The green photosynthetic carpels of heliotropic snow buttercup *(Ranunculus adoneus)* flowers contribute to the carbon balance for flower maintenance but at a potential cost in water loss. Although buttercup petals are characteristically waxy and resistant to water loss, the carpels are covered with functional stomata (Galen *et al.*, 1993). I used a simple potometer experiment to compare water loss from flowers oriented towards the sun (tracking, $n = 19$) versus flowers oriented directly upwards to simulate nontracking behavior ($n = 20$). Stems of cut flowers were inserted snugly into holes drilled in the lids of water-filled microcentrifuge tubes. Tubes held flowers at their natural height above the ground in the plant's alpine snowbed habitat (Pennsylvania Mountain, CO). Flowers of similar diameter were collected daily from 18–20 June 2000 and assigned at random to each treatment. The position of tracking flowers was adjusted hourly to maintain solar orientation and the amount of water remaining in the microcentrifuge tubes measured after 3–4 h. On average, tracking flowers lost more water per hour (0.034 ± 0.002 ml (SE) than stationary ones (0.025 ± 0.002 ml; mixed model ANOVA $F_{1,28} = 8.65$, $P < 0.0065$, Galen, unpublished data). Flowering in *R. adoneus* coincides with snowmelt, a time of surplus moisture availability to the vegetative plant. Consequently, heliotropic flowers, while luxurious in water use, are unlikely to negatively impact the water balance of the vegetative organs under extant climate/snow pack regimes.

A second experiment was done to determine how evapotranspiration by heliotropic flowers affects intrafloral temperature in *R. adoneus*. Under cool alpine conditions, solar tracking turns the parabolic snow buttercup flowers into "furnaces" by focusing sunlight on the inner whorls of sexual organs (Stanton and Galen, 1989). Heat captured by tracking flowers may promote rapid ovule growth, pollen germination, and pollinator visitation in the cold alpine environment (Galen and Stanton, 2003), but could reduce pollen and ovule viability under warmer conditions associated with future climate change. To address whether evapotranspiration might buffer heliotropic flowers from thermal stress, I measured the relationship between flower transpiration rate and internal flower temperature over an ambient temperature gradient. Snow buttercup flowers were measured in two contrasting orientations: towards the sun (to simulate tracking) and directly upward towards the zenith (to simulate stationary behavior). Measurements were taken by enclosing individual flowers in a modified "Conifer chamber" (Li-Cor 6400-05) attached to an infrared gas analyzer (IRGA, Li-Cor 6400, Fig. 3.3). Stems bearing fully expanded male phase flowers (to standardize flower age)

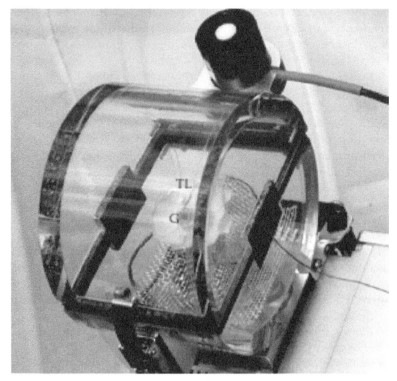

Figure 3.3 Modified "Conifer chamber" (Li-Cor Inc., 6400-05) adapted for measurements of gas exchange and intrafloral temperature in flowers of the snow buttercup, *Ranunculus adoneus*. TL indicates thermocouple lead to gynoecia (G). (See color plate.)

were excised and inserted individually through snugly fitting holes drilled into the lids of water-filled microcentrifuge tubes. The microcentrifuge tube was held in place by pushing it through a screen grid fit across the middle of the sampling chamber. Air temperature around the flower was controlled by the IRGA. An additional thermocouple was placed in contact with the carpels at the center of the flower to monitor flower temperature. Flowers exhibited a mean solar deviation of $51 \pm 2°$ (SE) and a rate of light interception averaging 1208 ± 113 (SE) $\mu mol/m^2/s$ when oriented upward, versus 2048 ± 112 (SE) $\mu mol/m^2/s$ when directly facing the sun. To insure that the enzymatic machinery for photosynthesis was fully active, measurements were taken at each ambient temperature with the flower first facing the sun and next oriented upward. Air temperature in the chamber ranged from 12 to 30°C and was increased in 3°C increments for each flower. Flowers were allowed to equilibrate at each temperature for at least 5 min before the measurement was logged. Flowers from 12 plants were subjected to a total of 83 observations (half in each orientation) between 0800–1700 h on clear days in July–August 2001–2004.

Results corroborate measurements of total water loss from flowers in the field. Solar tracking increased flower transpiration rate (E) by 8% on average (mixed model ANOVA, $F_{1,84} = 5.05$, $P < 0.029$). Internal flower temperature increased with air temperature in both tracking flowers and stationary flowers (regression, $P < 0.0001$ for both, Table 3.1, Fig. 3.4, Galen, unpublished data). However, the degree to which floral organs warmed with air temperature was buffered by transpirational cooling in tracking flowers (regression, $\beta = -1.57$, $P < 0.015$, Fig. 3.4) but not in stationary flowers, ($\beta = -0.56$, $P > 0.60$; Fig. 3.4). This result indicates that heliotropism stabilizes flower microclimate, elevating intrafloral temperature under cold alpine conditions, but cooling flowers under hot ambient conditions by promoting transpiration.

Although it is unlikely that temperature buffering by transpirational cooling is advantageous under currently cool alpine conditions, this mechanism may allow alpine and arctic plant species with heliotropic flowers to resist thermal damage to sexual organs in the future as the climate warms. Indeed, results from simulated warming manipulations suggest that plants with heliotropic flowers have physiological mechanisms allowing them to tolerate and even thrive under modest increases in ambient temperature. Totland (1999) found enhanced seed number and seed size when open-top chambers (OTCs) were used to increase air temperature for plants in an alpine population of solar tracking *Ranunculus acris*. Since use of water to regulate flower temperature may be economical only when soil moisture is plentiful during flowering, it would be worthwhile to re-examine the impact of global warming on fecundity in *Ranunculus* over an experimental moisture gradient.

Table 3.1 Multiple Regression of Intrafloral Temperature on Microclimate Parameters for Flowers of *Ranunculus adoneus* Oriented Towards the Sun (tracking) or Directly Upward (Stationary) (Galen, Unpublished Data)

	Solar tracking			Stationary		
Effect	Parameter estimate	t	$P<$	Parameter estimate	t	$P<$
Intercept	5.9	2.73	0.01	0.6	0.26	ns
Air temperature (C)	0.82	12.01	0.0001	1.05	11.45	0.0001
Transpiration Rate (E)	−1.57	2.54	0.015	−0.56	0.52	ns
Photosynthetically active radiation (PAR)	0.0021	2.00	0.052	0.0005	0.60	ns

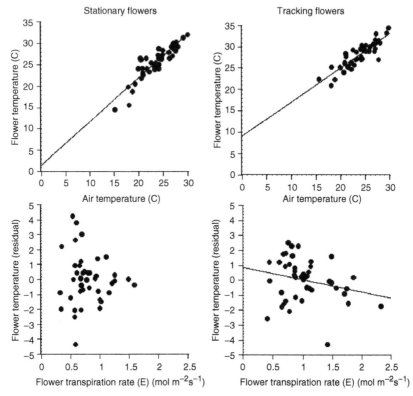

Figure 3.4 Intrafloral temperature as a function of air temperature (top) and as affected by transpirational cooling (bottom) in *Ranunculus adoneus* flowers oriented either directly upward (stationary) or towards the sun (tracking, Galen, unpublished data). Bottom plots show residual variation in flower temperature after effects of ambient temperature are accounted for through linear regression analysis. In both treatments, intrafloral temperature increases significantly with ambient temperature (Table 3.1). Flower transpiration has no effect on temperature of stationary flowers, but has a significant negative effect on temperature of tracking flowers (Table 3.1).

C. Plastic Responses of Floral Traits to Water Availability: Impact on Plant–Pollinator Interactions

A plant population's capacity to sustain its mutualist pollinators depends on the phenotypic distribution of floral rewards. Especially for larger pollinators with high metabolic demands, reductions in nectar resources can have devastating effects on local populations. Since many determinants of nectar production are sensitive to water availability, drought has a profound indirect effect on local pollinator abundance. For example, a severe drought at Finca La Selva (Costa Rica) in 1973 reduced flower abundance at the peak of the breeding season for the Long-tailed Hermit hummingbird, *Phaethornis superciliosus*, causing a 66% reduction in hummingbird population size the following year (Stiles, 1979, 1992). In the high alpine, years of extreme drought are associated with deficits in pollinator service for bumblebee pollinated *P. viscosum* (Fig. 3.5, Galen, unpublished data). It appears that in very disparate habitats, drought has similar negative effects on pollinator services, potentially leading to increased competition among plants for visits from the few remaining pollinators and enhancing the strength of pollinator-mediated selection on floral traits.

Water availability can also alter floral attractiveness to pollinators in a taxon-specific manner, causing shifts in the composition of the pollinator pool. For example, mason bees *(Chalicodoma sicula)* collect nectar from *Lotus creticus* as a source of water as well as energy and prefer dilute nectar over more concentrated nectar (Willmer, 1986). Nectar sugar concentration is one of the many examples of floral traits that are sensitive to water availability (Bertsche, 1983). It follows that spatial mosaics of soil moisture may create patches of plants that are differentially attractive to mason bees, with potential consequences for gene flow and local selection on floral traits in *L. creticus*. Soil moisture gradients can also cause pollinators to bypass plants in arid sites for more rewarding plants in mesic sites (Zimmerman, 1983). If habitat selection has a genetic component (as in dioecious species, where males often occupy drier sites than females, Dawson and Geber, 1999), then pollinators choosing among plants according to nectar production may indirectly exert selection on physiological traits that determine drought tolerance.

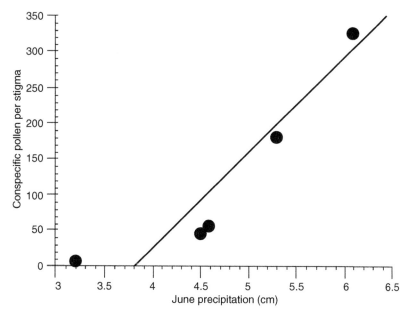

Figure 3.5   Relationship between average pollen receipt by flowers of bumblebee-pollinated *Polemonium viscosum* open in July on Pennsylvania Mountain, CO (means based on 15–185 plants per year) and total June precipitation (year 1985–1987, 2000, 2002, and 2004; NCDC climate monitoring, Colorado climate summary data), for the best-fit line, $r = 0.89$, $P < 0.0015$ (Galen, unpublished data).

For self-compatible species, changes in flower size or shape caused by moisture availability can directly affect outcrossing rates. Plasticity in morphological and developmental characters in relation to water availability determines the spatial or temporal juxtaposition of male and female sexual organs (e.g., Waller, 1980). In support of this view, herkogamy (spatial separation of anthers and stigma) is often reduced in hot, dry habitats (reviewed in Elle and Hare, 2002). Little is known about whether such plasticity in mating systems traits is advantageous. The argument could be made that reduced herkogamy under drought represents a mechanism for reproductive assurance under conditions of pollinator scarcity. To address this idea, experimental watering regimes could be used to create more or less herkogamous phenotypes that could then be placed into wet and dry environments. In this way, the match between floral phenotype and environment could be manipulated reciprocally to directly test the adaptive plasticity hypothesis (e.g., Schmitt *et al.*, 1999).

Perhaps the most common way that moisture regimes influence plant–pollinator interactions is by affecting the duration and timing of flower display (e.g., Minter and Lord, 1983; Corbet, 1990; Asmussen, 1993; Arathi *et al.*, 2002; Holland, 2002; Holland and Fleming, 2002). Drought stress delays flowering of senita cacti (*Lophocereus schottii*), causing flowers to open under warm summer conditions instead of cooler spring temperatures. Anthesis in senita flowers begins at sunset, continues overnight, and ends as water loss causes the corolla to wilt the next day. Temperature advances the onset of wilting, leading to shorter longevity for summer flowers than spring flowers. Flowers opening in the summer are moth-pollinated, because wilting occurs before daybreak. Conversely, flowers opening in the spring last longer and are serviced by copollinating bees as well as moths. This example beautifully illustrates how water availability can influence the complement of pollinators servicing a host plant species, and in so doing, determine the degree of specialization in the plant–pollinator interaction.

IV. Conclusions

In this chapter, I have made the case that water fulfills a diversity of functions in the life of a flower and has both positive and negative impacts on plant mating success. A complete treatment of water as a key player in the process of sexual reproduction may begin with resource limitation, but must also consider parental environmental effects and conduit effects related to the physical and biochemical properties of water. More research on the physiology of drought stress in flowers is needed, especially in the realm of hormonal controls. Nonetheless, we know much more about the ecophysiology of flower responses to drought stress than about the implications of such responses for floral evolution. In part, our lack of knowledge stems from the tradition of pollination ecologists to focus on pollinators as the dominant mode of selection on floral traits (reviewed in Galen, 1999). This habit is slowly changing, as researchers recognize that the abiotic context in which pollination occurs is more than a source of noise making pollinator-mediated selection difficult to detect. Environmental stress can exaggerate costs and reduce the benefits of pollination mutualisms, disrupting specialized relationships and favoring generalization or even selfing. Ignoring stress factors may limit our understanding of why most flowering plants exhibit generalized pollination modes and mixed mating systems.

Model organisms (e.g., *Arabidopsis, Brassica rapa, Helianthus annuus*) and their wild relatives have much to offer for studies of physiological constraints on floral trait expression and pollinator attractiveness. For example, it is possible to design experiments in which floral traits and pollination success are measured as a function of specific changes in gene expression under drought stress. Consider the plastid lipid-associated gene (PAP-2) in *Brassica rapa*. This gene encodes carotenoid pigments in the petals that influence attractiveness to insect pollinators. Expression of PAP-2 is down regulated under drought (Kim *et al.*, 2001). Experiments comparing gene expression at the PAP-2 locus, petal color intensity, and pollinator visitation for plants under wet and dry conditions could directly test the role of drought stress in moderating plant–pollinator interactions.

Advances in technology also make the measurement of physiological traits associated with drought stress or tolerance a straightforward task even in remote habitats. Gas exchange measurements can be used to measure instantaneous responses of plants to drought stress (e.g., Heschel *et al.*, 2002). Additionally, carbon isotope analysis can be applied to measure integrated long-term responses to drought (e.g., Ward *et al.*, 2002). Together, these technologies facilitate the measurement of physiological costs associated with water use by flowers over diverse time scales relevant to growth and survival. Simultaneous measurements of plant fitness components should enable the evolutionary consequences of physiological costs to be addressed as well.

In conclusion, much remains to be done if we are to understand how water relations influence pollination, mating success and, by extension, floral evolution. New technologies and ways of thinking about plant pollination make this a tractable goal. Longstanding threats associated with pollution, habitat degradation, and climate change make it an important one – not only for agriculture but for the conservation of wildflowers and the animals whose livelihoods depend upon them, over the millennia to come.

Acknowledgments

I thank Marc Brock, Amy Dona Leah Dudley, Joseph Riley, and an anonymous reviewer for comments on the manuscript; Robert Sharp, Stephen Pallardy, and Bill Davies for helpful discussions, Lynda Delph and Todd

Dawson for help in adapting the Li-Cor 6400 for flower measurements, and Devon Pattillo for assistance in the field. Funding was provided by the University of Missouri Research Board and MU Research Council, and by NSF grants BSR 8604726, DEB 0087412 and DEB 0316110.

References

Andersson S. (2000) The cost of flowers in *Nigella degenii* inferred from flower and perianth removal experiments. *Int J Plant Sci* **161**: 903–908.

Arathi H. S., Rasch A., Cox C. and Kelly J. K. (2002) Autogamy and floral longevity in *Mimulus guttatus. Int J Plant Sci* **163**: 567–573.

Armstrong J. E. (2002) Fringe science: are the corollas of *Nymphoides* (Menyanthaceae) flowers adapted for surface tension interactions. *Am J Bot* **89**: 362–365.

Ashman T. L. and Schoen D. J. (1994) How long should flowers live? *Nature* **371**: 788–790.

Asmussen C. B. (1993) Pollination biology of the sea pea *Lathyrus japonicus*: floral characters and activity and flight patterns of bumblebees. *Flora* **188**: 227–237.

Barrett S. C. H. (1992) Gender variation and the evolution of dioecy in *Wurmbea dioica* (Liliaceae) *J Evol Biol* **4**: 423–444.

Bertsche A. (1983) Nectar production of *Epilobium angustifolium* at different air humidities. Nectar sugar in individual flowers and the optimal foraging theory. *Oecologia* **59**: 40–48.

Boose D. L. (1997) Sources of variation in floral nectar production rate in *Epilobium canum* (Onagraceae): implications for natural selection. *Oecologia* **110**: 493–500.

Boot R., Raynal D. J. and Grime J. P. (1986) A comparative study of the influence of drought stress on flowering in *Urtica dioica* and *Urtica urens. J Ecol* **74**: 485–496.

Bynum M. R. and Smith W. K. (2001) Floral movements in response to thunderstorms improve reproductive effort in the alpine species *Gentiana algida* (Gentianaceae) *Am J Bot* **88**: 1088–1095.

Carroll A. B., Pallardy S. G. and Galen C. (2001) Drought stress, plant water status and floral trait expression in fireweed, *Epilobium angustifolium* (Onagraceae) *Am J Bot* **88**: 438–446.

Case A. L. and Barrett S. C. H. (2001) Ecological differentiation of combined and separate sexes of *Wurmbea dioica* (Colchicaceae) in sympatry. *Ecology* **82**: 2601–2616.

Catling P. M. (1980) Rain assisted autogamy in *Liparis–loeselii*, Orchidaceae. *Bull Torr Bot Club* **104**: 525–529.

Ceccarelli S. and Grando S. (1996) Drought as a challenge for the plant breeder. *Plant Growth Reg* **20**: 149–155.

Chiarello N. R. and Gulmon S. L. (1991) Stress effects on plant reproduction. In *Responses of Plants to Multiple Stresses* (H. A. Mooney, W. E. Winner and E. J. Pell eds.) pp. 162–188, Academic Press, New York.

Corbet S. A. (1990) Pollination and the weather. *Israel J Bot* **39**: 13–30.

Costich D. E. (1995) Gender specialization across a climatic gradient: experimental comparison of monoecious and dioecious *Ecballium Ecology* **76**: 1036–1050.

Cox R. M. (1984) Sensitivity of forest plant reproduction to long range transported air pollutants: *in vitro* sensitivity of pollen to simulated acid rain. *New Phytol* **97**: 63–70.

Cushman J. C. (2001) Osmoregulation in plants: implications for agriculture. *Am Zool* **41**: 758–769.

Davies W. J., Bacon M. A., Thompson D. S., Sobeih W. and González L. (2000) Regulation of leaf and fruit growth in plants growing in drying soil: exploitation of the plants' chemical signaling system and hydraulic architecture to increase the efficiency of water use in agriculture. *J Exp Bot* **51**: 1617–1626.

Dawson T. E. and Geber M. A. (1999) Sexual dimorphism in physiology and morphology. In *Gender and Sexual Dimorphism in Flowering Plants* (M. A. Geber, T. E. Dawson and L. F. Delph eds.) pp. 175–216, Springer, Berlin.

De la Berra E. and Nobel P. S. (2004) Nectar: properties, floral aspects and speculations on origin. *Trends Plant Sci* **9**: 65–69.

Dodd J. C., Estrada A. B., Matcham J., Jeffries P. and Jeger M. J. (1991) The effect of climatic factors on *Colletotrichum-gloeosporioides* causal agent of mango anthracnose in the Phillipines. *Plant Path* **40**: 568–575.

Eisokowitch D. and Woodell S. R. J. (1974) The effect of water on pollen germination in two species of *Primula*. *Evolution* **28**: 692–694.

Elle E. and Hare J. D. (2002) Environmentally induced variation in floral traits affects the mating system in *Datura wrightii*. *Funct Ecol* **16**: 79–88.

Faegri K. and van der Pijl L. (1971) The principles of pollination ecology. Oxford: Pergamon.

Galen C. (1989) Measuring pollinator-mediated selection on morphometric floral traits: bumblebees and the alpine skypilot, *Polemonium viscosum*. *Evolution* **43**: 882–890.

Galen C. (1996) Rates of floral evolution: adaptation to bumblebee pollination in an alpine wildflower, *Polemonium viscosum*. *Evolution* **46**: 1043–1051.

Galen C. (1999) Why do flowers vary? The functional ecology of variation in flower size and form within natural plant populations. *BioScience* **49**: 631–640.

Galen C. (2000) High and dry: drought stress, sex-allocation trade-offs and selection on flower size in the alpine wildflower, *Polemonium viscosum* (Polemoniaceae) *Am Nat* **156**: 72–83.

Galen C., Zimmer K. A. and Newport M. E. (1987) Pollination in floral scent morphs of *Polemonium viscosum*: a mechanism for disruptive selection on flower size. *Evolution* **41**: 599–606.

Galen C., Dawson T. E. and Stanton M. L. (1993) Carpels as leaves: meeting the carbon cost of reproduction in an alpine buttercup. *Oecologia* **95**: 187–193.

Galen C., Sherry R. A. and Carroll A. B. (1999) Are flowers physiological sinks or faucets? Costs and correlates of water use by flowers of *Polemonium viscosum*. *Oecologia* **118**: 461–470.

Galen C. and Stanton M. L.(2003) Sunny-side up: flower heliotropism as a source of parental environmental effects on pollen quality and performance in the snow buttercup, *Ranunculus adoneus* (Ranunculaceae) *Am J Bot* **90**: 724–729.

Grant B. R. (1985) Selection on bill characters in a population of Darwin's finches: *Geospiza conirostris* on Isla Genovesa, Galapogos. *Evolution* **39**: 523–532.

Guerrant E. O. (1984) The role of ontogeny in the evolution and ecology of selected species of *Delphinium* and *Limnanthes*. Unpublished doctoral dissertation, University of California, Berkeley.

Gilissen L. J. W. (1977) The influence of relative humidity on the swelling of pollen grains in vitro. *Planta* **137**: 299–301.

Guinn G., Dunlap J. R. and Brummett D. L. (1990) Influence of water deficits on the abscisic acid and IAA contents of flower buds and flowers. *Plant Physiol.* **93**: 1117–1120.

Hashem A., Majumdar M. N. A., Hamid A. and Hossain M. M. (1998) Drought stress effects on seed yield, yield attributes, growth, cell membrane stability and gas exchange of synthesized *Brassica napus* L. *J Agron Crop Sci* **180**: 129–136.

Heschel M. S., Donohue K., Hausmann N. and Schmitt J. (2002) Population differentiation and natural selection for water-use efficiency in *Impatiens capensis* (Balsaminaceae). *Int J Plant Sci* **163**: 907–912.

Hewett E. W. and Young K. (1980) Water sprinkling to delay bloom in fruit trees. *New Zeal J Agric Res* **23**: 523–528.

Hildebrand P. D., McRae K. B. and Lu X. (2001) Factors affecting flower infection and disease severity of lowbush blueberry by *Botrytis cinera*. *Can J Plant Path* **23**: 364–370.

Holland J. N. (2002)Benefits and costs of mutualism: demographic consequences in a pollinating seed-consumer interaction. *Proc R Soc B* **269**: 1405–1412.

Holland J. N. and Fleming T. H. (2002) Co-pollinators and specialization in the pollinating seed-consumer mutualism between senita cacti and senita moths. *Oecologia* **133**: 534–540.

Huang S.Q., Takahashi Y. and Dafni A. (2002) Why does the flower stalk of *Pulsatilla cernua* (Ranunculaceae) bend during anthesis. *Am J Bot* **89**: 1599–1603.

Jaafar H. Z., Atherton J. G., Black C. R. and Roberts J. A. (1998) Impact of water stress on reproductive development of sweet peppers (*Capsicum annuum* L.). I. Role of ethylene in water deficit-induced flower abscission. *J Trop Agric and Food Sci* **26**: 165–174.

Jonas C. S. and Geber M. A. (1999) Variation among populations of *Clarkia unguiculata* (Onagraceae) along altitudinal and latitudinal gradients. *Am J Bot* **86**: 333–343.

Kim H. U., Sherry S. H., Wu C. R. and Huang A. H. C. (2001) *Brassica rapa* has three genes that encode proteins associated with different neutral lipids in plastids of specific tissues. *Plant Physiol* **126**: 330–341.

Kokubun M., Shimada S. and Takahashi M. (2001) Flower abortion caused by preanthesis water deficit is not attributed to impairment of pollen in soybean. *Crop Sci* **41**: 1517–1521.

Laporte M. M. and Delph L. F. (1996) Sex-specific physiology and source-sink relations in the dioecious plant *Silene latifolia. Oecologia* **106**: 63–72.

LePage-Degivry M. T., Orlandini M., Garello G., Barthe P. and Gudin S. (1991) Regulation of ABA levels in senescing petals of rose flowers. *J Plant Growth Reg* **10**: 67–72.

Levin D. A. and Watkins L. (1984) Assortative mating in *Phlox. Heredity* **53**: 595–602.

Lisci M., Tanda C. and Pacini E. (1994) Pollination ecophysiology of *Mercurialis annua* L. (Euphorbiaceae), an anemophilous species flowering year round. *Ann Bot* **74**: 125–135.

Lord E. (1980) Physiological controls on the production of cleistogamous and chasmogamous flowers in *Lamium amplexicaule. Ann Bot* **44**: 757–766.

Lush W. M., Grieser F. and Wolters-Arts M. (1998) Directional guidance of *Nicotiana alata* pollen tubes in vitro and on the stigma. *Plant Physiol* **118**: 733–741.

Minter T. C. and Lord E. M. (1983) Effects of water stress abscisic acid and gibberellic acid on flower production and differentiation in the cleistogamous species *Collomia grandiflora* (Polemoniaceae) *Am J Bot* **70**: 618–624.

Mohan Ram H. Y. and Rao I. V. (1984) Physiology of flower bud growth and opening. *Proc Ind Acad Sci* **93**: 253–274.

Monson R. K., Grant M. C., Jaeger C. H. and Schoettle A. W. (1992) Morphological causes for the retention of precipitation in the crowns of alpine plants. *Env Exp Bot* **32**: 319–327.

Mouhouche B., Rouget F. and Delecolle R. (1998) Effects of water stress applied at different phenological phases on yield components of dwarf bean (*Phaseolus vulgaris* L.). *Agronomie* **18**: 197–205.

Nam N. H., Chauhan Y. S. and Johansen C. (2001) Effect of timing of drought stress on growth and grain yield of extra-short duration pigeon pea lines. *J Agric Sci* **136**: 179–189.

Nobel P. S. (1977) Water relations of flowering in *Agave deserti. Bot Gaz* **138**: 1–6.

Panavas T., Walker E. L. and Rubinstein B. (1998) Possible involvement of abscisic acid in senescence of daylily petals. *J Exp Bot* **49**: 1987–1997.

Passioura J. B. (1996) Drought and drought tolerance. *Plant Growth Reg* **20**: 79–83.

Patiño S. and Grace J. (2002) The cooling of convolvulaceous flowers in a tropical environment. *Plant, Cell Environ* **25**: 41–51.

Pusey P. L. (2000) The role of water in epiphytic colonization and infection of pomaceous flowers by *Erwinia amylovora. Phytopathology* **90**: 1352–1357.

Schlaepfer M. A., Runge M. C. and Sherman P. W. (2002) Ecological and evolutionary traps. *Trends Ecol Evol* **17**: 474–479.

Schmitt J., Dudley S. A. and Pigliucci M. (1999) Manipulative approaches to testing adaptive plasticity: phytochrome-mediated shade avoidance responses in plants. *Am Nat* **154**: S43–S54.

Sharp R. E. and LeNoble M. E. (2002) ABA, ethylene and the control of shoot and root growth under water stress. *J Exp Bot* **53**: 33–37.

Sheviak C. J. (2001) A role for water droplets in the pollination of *Platanthera aquilonis* (Orchidaceae) *Rhodora* **103**: 380–386.

Sklenar P. (1999) Nodding capitula in the superparamo Asteraceae: an adaptation to an unpredictable environment. *Biotropica* **31**: 394–402.

Smith-Huerta N. L. and Vasek F. C. (1987) Effects of environmental stress on components of reproduction in *Clarkia unguiculata*. *Am J Bot* **74**: 1–8.

Stanton M. L. and Galen C. (1989) Consequences of flower heliotropism for reproduction in an alpine buttercup (*Ranunculus adoneus*) *Oecologia* **78**: 477–485.

Stiles F. G. (1979) The annual cycle of a co-adapted community of hummingbirds and flowers in a tropical wet forest of Costa Rica. *Revista Biologia Tropical* **27**: 75–102.

Stiles F. G. (1992) Effects of a severe drought on the population biology of a tropical hummingbird. *Ecology* **73**: 1375–1390.

Totland O. (1999) Effects of temperature on performance and phenotypic selection on plant traits in alpine *Ranunculus acris*. *Oecologia* **120**: 242–251.

Turner L. B. (1993) The effect of water stress on floral characters, pollination and seed set in white clover (*Trifolium repens* L.). *J Exp Bot* **44**: 1155–1160.

Waller D. M. (1980) Environmental determinants of outcrossing in *Impatiens-capensis* Balsaminaceae. *Evolution* **34**: 747–761.

Ward J. K., Dawson T. E. and Ehleringer J. R. (2002) Responses of *Acer negundo* genders to interannual differences in water availability determined from carbon isotope ratios of tree ring cellulose. *Tree Physiol* **22**: 339–346.

Webb C. J. (1999) Empirical studies: evolution and maintenance of dimorphic breeding systems. In *Gender and Sexual Dimorphism in Flowering Plants* (Geber M. A., Dawson T. E. and Delph L. F. eds.) pp. 61–122, Springer, Berlin.

Weller S. G., Wagner W. L. and Sakai A. K. (1995) A phylogenetic analysis of *Schiedea* and *Alsinidendron* (Caryophyllaceae: Alsinoideae): implications for the evolution of breeding systems. *Syst Bot* **20**: 315–337.

Westgate M. E. and Peterson C. M. (1993) Flower and pod development in water-deficient soybeans (*Glycine max* L. Merr.). *J Exp Bot* **44**: 109–117.

Whiley A. W., Chapman K. R. and Saranah J. B. (1988) Water loss by floral structures of Avocado (*Persea americana* cv. Fuerte) during flowering. *Aust J Agric Res* **39**: 457–467.

Willmer P. G. (1986) Foraging patterns and water balance problems of optimization for a xerophilic bee *Chalicodoma-sicula*. *J Anim Ecol* **55**: 941–962.

Willmer P. G. (1988) The role of insect water balance in pollination ecology: *Xylocopa* and *Caloptropis*. *Oecologia* **76**: 430–438.

Yamane K., Kawabata S. and Sakiyama R. (1991) Changes in water relations, carbohydrate contents and acid invertase activity associated with perianth elongation during anthesis of cut gladiolus flowers. *Journal of the Japan Soc Hort Sci* **60**: 421–428.

Zimmerman M. (1983) Plant reproduction and optimal foraging: experimental nectar manipulations in *Delphinium nelsonii*. *Oikos* **41**: 57–63.

Zimmerman M. and Pyke G. H. (1988) Experimental manipulations of *Polemonium foliosissimum*: effects on subsequent nectar production, seed production and growth. *J Ecol* **76**: 777–789.

Zuberi M.I. and Dickinson H.G. (1985) Pollen-stigma interactions in Brassica. III. Hydration of the pollen grains. *J Cell Sci* **76**: 321–336.

4

The Allometry of Reproductive Allocation

Gregory P. Cheplick

I. Introduction

The concept of allometry has a long and convoluted history, predominantly applied to animals by zoologists, such as W. D'Arcy Thompson, J. S. Huxley, and more recently, S. J. Gould (Thompson, 1917; Huxley, 1932; Gould, 1966; Gayon, 2000). The term derives from the Greek words for measure *(metron)* and other *(allo)*, referring to the changes in one metric trait (e.g., organ size) relative to that of a second trait (e.g., body mass). Often allometric analysis involves simple scaling principles used to examine size-correlated variation in organ number, size, shape, or physiology (Reiss, 1989). Although comparative animal physiology has a tradition of examining the relation of aspects of physiology such as oxygen consumption to body size (Peters, 1983), allometric studies of plants have historically focused on morphological traits such as biomass. While general trends in allometry have been difficult to develop for plants, the growth rates of both animals and plants are surprisingly similar, as described by Damuth (2001). Many of the general scaling principles important to plants have been explored in detail by Niklas (1994) in his text on *Plant Allometry*.

Using a dynamic, developmental perspective, the standard dictionary definition of allometry is the growth of one part of an organism relative to the growth of the entire organism, or some other part of it (Niklas, 1994). This definition implies changes accruing along a growth trajectory; in actual practice, many researchers examine *static* allometry, correlations among individuals at a specific ontogenetic stage (Schlichting and Pigliucci, 1998). Examples of static allometric relationships in animals include positive correlations between brain size and body mass among primate species (Gould, 1975), and between leg length and body size for a species of water strider (Tseng and Rowe, 1999). Note from these examples that both interspecific and intraspecific allometric relationships can be developed (Reiss, 1989; Gayon, 2000; Niklas and Enquist, 2002a,b).

Since the shape of an organism is greatly determined by the relative growth rates of its different parts over development, changes in allometric growth can modify adult body form. Whenever there exists genetic variation for the relative growth rates of component parts, natural selection may modify the allometric relations within a population, presumably in such a way that fitness is maximized. In short, allometric relationships can be considered an evolved component of a species life history strategy (Schlichting and Pigliucci, 1998).

Over the past several decades, plant scientists have become increasingly interested in allometry and its relation to ecological and evolutionary processes (Weiner, 1988a; Niklas, 1994; Zhang and Jiang, 2002). This is partly because plant size seems to be important for many aspects of ecology and evolution, including population dynamics and growth, interspecific and intraspecific competition, and life history strategies. One of the major consequences of environmental heterogeneity is widespread variation in the size and reproduction of individuals (Weiner, 1988b; Stewart *et al.*, 2000). The myriad of abiotic and biotic factors that impinge upon plant populations influence reproductive output, allocation, and evolutionary fitness. For herbaceous species, shifts in allocation patterns in response to changing environmental conditions are presumed to maximize growth and reproduction (Bazzaz, 1996, 1997). However, much of the variation in reproductive output and allocation found among individuals in natural populations and across experimental treatments is likely due to size-dependent effects. Hence, it is imperative to consider allometry when trying to disentangle the impacts of environmental factors on plant reproduction (Samson and Werk, 1986).

In this chapter the focus will be on reproductive allocation (RA) and its relation to plant size, assessed as vegetative dry mass. The objectives are to (1) describe how allometric theory can be applied to the concept of RA, (2) determine whether or not RA is correlated with Darwinian fitness by presenting data for a few annual and perennial species, (3) explore how development impacts the allometry of reproductive modules in some detail for an annual plant, (4) show how the allocation of biomass to a single propagule (here termed reproductive expenditure) may be useful in allometric research because it is commonly negatively related to plant size, (5) review the allometry literature to determine whether or not any generalizations can be made regarding plant life history and the allometry of RA, and (6) use path analysis to explore the functional components of allometry by examination of the inter-relationships among root, stem, and leaf mass and RA, and the indirect effects of other traits on vegetative mass.

II. Definition and Analysis of RA in Relation to Allometry

Much has been written about RA and its various definitions (for reviews, see Bazzaz and Ackerly, 1992; Bazzaz, 1997; Reekie, 1999). In this chapter, RA will be expressed as the dry mass of the sexual reproductive structures (flowers, fruits, seeds) relative to vegetative mass (shoots plus roots, if root data are available). As defined by Bazzaz and Ackerly (1992), RA is the "proportion of the total resource supply devoted to reproductive structures." As such, RA differs from "reproductive effort" which refers to the investment of a resource in reproduction that has resulted in its diversion from vegetative activities (see Bazzaz and Ackerly, 1992, for further discussion).

Most workers agree that, to avoid spurious correlation, for statistical analysis the denominator of the ratio used to calculate RA should not be the total summed mass of vegetative and reproductive parts (Samson and Werk, 1986; Wagner, 1989; Klinkhamer *et al.*, 1992; Reekie, 1998; Cheplick, 2001). Analysis of the allometry of reproductive mass in relation to vegetative mass is straightforward, because the assumption of regression analysis that variables are independent is met. Unfortunately, for the allometric analysis of RA to plant size, the problem of spurious correlation remains, because RA is defined based on a measure of size such as vegetative mass. In other words, the assumption of complete independence of variables is violated when regressing RA on vegetative mass, as well as total mass. This problem has been recognized before and is not generally considered to be significant whenever reproductive mass is relatively low compared to vegetative mass, as in many perennials (Samson and Werk, 1986; Niklas, 1994; Niklas and Enquist, 2003). Presently there are no formal statistical solutions that have been developed for dealing with these problems in analyzing the allometry of RA. Hence, all regression models developed here and elsewhere for the relation of RA to size must be viewed with caution, especially with regard to comparisons across species or experimental treatments. However, when comparing groups that span a similar range in vegetative mass, it is possible to examine variation in the regression parameters to reach general conclusions regarding allometric relationships.

In the analysis of ecological data, regression analysis can be combined with an analysis of variance in an ANCOVA to ascertain if treatment differences are due mostly to different sizes of organisms across treatments. Thus, when plant size is entered as a covariable into an analysis of covariance, the effect of an experimental treatment on RA can be assessed while statistically controlling for size, if that is what the investigator wishes to do. If after performing an ANCOVA, the putative treatment effects on RA all disappear, then the investigator can surmise that the treatments changed plant size which directly affected RA. The investigator should not lament this result – it simply reinforces how important size is to plant populations and why allometry is relevant. Despite the caveats, least squares or reduced major axis regression alone, or when incorporated into an ANCOVA, remains a powerful tool necessary to elucidate an allometric relationship of RA to size, if one exists.

As an example of how allometry and the definition of RA can be relevant to the interpretation of reproductive patterns, consider the changes in reproductive allocation reported to occur across habitats that differ in successional age (Hancock and Pritts, 1987). For the herbaceous weed *Potentilla recta*, Soule and Werner (1981) detected a significant decline in RA from the youngest (10 year) to the intermediate (15 year) to the most mature (25 year) habitats in the first year of a two-year study. A plot of RA (reproductive mass/vegetative mass) versus vegetative mass reveals that the differences in RA among plants in different aged fields were mostly due to the smaller size of plants in older fields (Fig. 4.1). Soule and Werner (1981), while calculating RA as the ratio of reproductive mass to

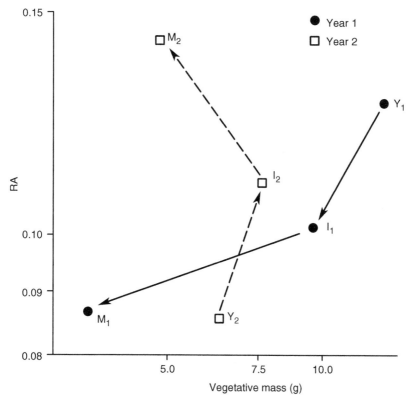

Figure 4.1 Contrasting patterns in the relation of reproductive allocation (RA) to vegetative mass for *Potentilla recta* collected in two years from three successional habitats: young (Y), intermediate (I), and mature (M). Data formulated from Table 2 of Soule and Werner (1981).

aboveground mass, reported no differences in RA or biomass among fields in the second year of the study. However, when RA was reformulated as the ratio of reproductive to total vegetative mass (i.e., aboveground mass *plus* roots), and plotted allometrically, a striking trend emerged for the second-year data. In contrast to the first year, and despite relatively low variation in vegetative mass among fields, RA showed a major *increase* from the youngest to the oldest habitat (Fig. 4.1). Clearly, when attempting to unravel ecological trends in RA, consideration should be given to both the definition of RA used and its allometric relationship to plant size.

III. Allometry Theory and RA

How can allometry theory contribute to our understanding of the many patterns of reproductive allocation observed in plants? Previous studies have used allometric equations to relate density to mean plant mass ("self-thinning"; White and Harper, 1970; White, 1981), the height of the trees to the diameter of their trunks (Whittaker and Woodwell, 1968), vegetative growth to organ mass (Niklas and Enquist, 2002a,b), and seed output to seed mass (Kawano, 1981). The well-known power function $Y = \beta X^{\alpha}$ (Huxley, 1932) is useful for many allometric relationships, where β is the Y-intercept (scaling coefficient), α is the slope (scaling exponent), and X and Y are the independent and dependent variables, respectively. For many intraspecific comparisons, a simple linear equation will give similar results, when Y is regressed against X and may be easier to interpret (Samson and Werk, 1986): $Y = \beta + \alpha X$, where β is the Y-intercept and α is the slope (Niklas, 1994). Least squares or reduced major axis regression provides the correlation coefficients used to assess the allometry between X and Y. The resulting regression "curve" will appear linear when the scaling relationship is depicted on a standard $\log_{10}$ versus $\log_{10}$ plot (Niklas, 1994).

Samson and Werk (1986) constructed a graphical model of size-dependent variation in the ratio of reproductive to vegetative mass, which they termed reproductive effort. Provided that there was a linear relationship between reproductive and vegetative mass, allocation to reproduction could decrease, increase, or remain the same with increasing vegetative mass. The nature of the relationship between reproductive allocation and size simply depended on whether $\beta > 0$ (RA decreases), $\beta < 0$ (RA increases), or $\beta = 0$ (RA is constant). Thus, developmental and structural constraints with increasing size could produce allometric relationships with size-dependent variation in proportional allocation. Since many differences in RA reported among habitats or experimental conditions could be due to size (e.g., Sadras *et al.*, 1997), Samson and Werk (1986) maintained that "every study of plant life history should test whether populations exhibit size-dependent reproductive-allocation effects." Clearly, the evolution of a particular pattern of RA in relation to environmental conditions may reflect selection for traits that are mostly relevant to plant size, but indirectly, may impact RA.

Hara *et al.* (1988) developed another RA model that assumed a plant could control its allocation so as to maximize its investment in reproduction at the end of a given growth period. They explored their model in light of field and experimental studies of species that varied in life form and habitat requirements. In woodland perennials, the larger the plant size, the smaller the RA. The model predicted this allometric relationship for most species common in closed, stable communities. Annuals and biennials from changing, unpredictable environments showed large variation in size, but a typically constant RA. Klinkhamer *et al.* (1992) later reported that nonlinear relationships between reproductive output and size could elicit a hump-shaped relationship between RA and size. Their analysis of data from three perennial species did not show a straightforward, monotonic increase or decrease in RA with size.

Clearly, the application of allometry theory to the principle of allocation is fraught with difficulty, given the diversity of plant life forms and morphological structures. Models of resource allocation to sexual reproduction in clonal plants capable of vegetative reproduction must consider the relative costs and benefits of producing sexual versus asexual propagules (Armstrong, 1982; Loehle, 1987; Bazzaz, 1997). From a practical standpoint, plant size may be difficult to define for large, perennial herbs that produce numerous ramets, and sexual reproduction may not necessarily be related to long-term fitness. For woody plants, size may not be meaningful for allometric research because of the massive investment in nonliving, supporting structures. Unfortunately, most models to date do not incorporate the costs of producing supporting structures (e.g., the rachis of an inflorescence) or the carbohydrate costs of storage organs (e.g., roots or corms).

Since RA, as defined earlier, already incorporates a measure of plant size (typically biomass), what is the biological meaning of the allometry of RA? As will be seen, many allocation studies have presented graphical correlations of RA with vegetative mass without considering this important question. Because RA is a measure of reproductive mass divided by vegetative mass, it is a unitless proportion. Ecologists have long recognized that RA is flexible in relation to environmental conditions (Bazzaz, 1997). RA can also change during ontogeny (e.g., Sadras *et al.*, 1997), and developmental constraints associated with increasing size can result in "variable proportional allocation with size" (Samson and Werk, 1986). The relationship of RA to plant size can help illuminate ecologically relevant aspects of life history strategies, because it reveals the partitioning of photosynthetically fixed carbon to functionally different tissues (Farrar and Gunn, 1998).

When reproductive mass changes in direct linear proportion to increases in vegetative mass, RA will not vary with vegetative mass (e.g., *Poa annua,*

Wagner, 1989). A decrease in RA with vegetative mass, however, indicates plants are partitioning an increasingly greater proportion of the resource pool to nonreproductive functions as they become larger. For example, if the relative balance of allocation to reproductive versus vegetative tissues shifts in favor of resource-acquiring organs or storage tissues, as plants grow larger, then RA will decline with increasing vegetative mass. This can occur even though absolute reproductive mass increases in relation to vegetative mass (Samson and Werk, 1986). Such a strategy could be adaptive in a closed, competitive environment, and might occur in clonal perennials that often show low allocation to sexual reproduction (see Section VI). Alternatively, a higher proportional allocation to sexual reproduction (e.g., seeds) at small sizes may indicate a strategy that is adaptive in an unstable, disturbed environment, where the risk of early death before reproduction is high. For a species (or experimental treatment) where RA increases with vegetative mass, plants are partitioning an increasingly greater proportion of the resource pool into sexual reproductive functions as they become larger. This pattern may represent an opportunistic strategy where maximizing fruit and seed production is the major component of Darwinian fitness (e.g., colonizing annuals). As allometry incorporates ontogenetic effects and is a product of the net fluxes of carbon occurring within the plant, it is thought to be related to the processes that underlie resource partitioning (Farrar and Gunn, 1998). The allometry of RA can therefore be a way to summarize and compare resource-partitioning patterns in relation to environmental or experimental conditions for many plant species.

A potential complicating factor that should be incorporated into allometry-based models of RA is the well-known trade-off between seed size and seed number. Theoretically, two plants may produce different number of seeds but show the same RA at a given size (when seed mass varies); the evolutionary and ecological significance of this possibility has not been fully explored.

One way to incorporate seed number into the description of the allometry of RA is to calculate allocation to a single seed by dividing RA by the number of seeds produced per individual. Kawano *et al.* (1989) referred to this parameter as the "cost of producing a single propagule." Here, the term "cost" will be avoided, because it has traditionally been used to depict the reduction in survival, growth, or future reproduction that typically occurs following an episode of reproduction (Reekie and Bazzaz, 1987, 1992; Reekie, 1999; Obeso, 2002). Instead, the parameter will be referred to as reproductive expenditure because it reflects the energy allocated to an individual seed. As might be predicted, given the trade-off between seed number and mass, reproductive expenditure decreases with increasing seed number and decreasing seed mass (Kawano, 1981).

To explore the potential utility of reproductive expenditure in the description of the allometry of RA, mean values of RA and expenditure were obtained from a detailed study of two annuals subjected to experimental treatments that varied either with density or nitrogen levels (Kawano *et al.,* 1989). The mean values per treatment were plotted against mean plant biomass on $\log_{10}$ versus $\log_{10}$ axes. For *Oryza sativa* cv. *Akihikari* grown at densities from 1 to over 4000 plants m^{-2}, individual mass varied 25-fold, but RA was remarkably constant (Fig. 4.2). For *Coix ma-yuen* grown at five levels of soil nitrogen, individual mass varied 9-fold; however, RA increased with mass up to intermediate nitrogen levels, but then leveled off at 30% at the highest nitrogen levels (Fig. 4.2). Although the allometric pattern for RA is different for the two species, a plot of expenditure versus mass revealed a tight negative relationship for both species (Fig. 4.2), suggesting that expenditure is sharply reduced for large plants despite different causes for the variation in size (due to the small number of mean values and unavailability of raw data, statistical analyses were not attempted). Later, the allometry of expenditure will be examined again for an annual and perennial grass that show contrasting patterns for the allometry of RA.

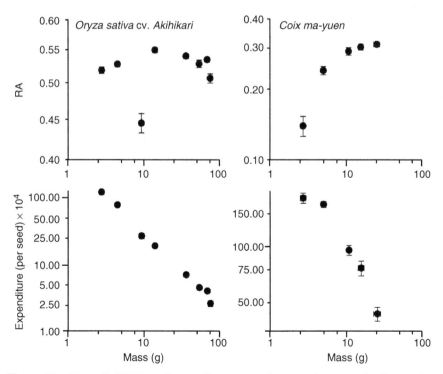

Figure 4.2 Mean (± SE) RA and expenditure per seed versus plant mass for *Oryza sativa* cv. *Akihikari* cultivated at densities from 1 to 4445 plants m^{-2} (left) and for *Coix ma-yuen* cultivated at five mineral fertilizer levels (right). Data are from Tables 1 and 3 for *O. sativa* and Table 6 for *C. ma-yuen* in Kawano *et al.* (1989).

IV. Relation of RA to Relative Fitness

In their detailed review of reproductive allocation in plants, Bazzaz and Ackerly (1992) maintained that "if two plants of equal size differ in their allocation to reproduction, the individual with the higher allocation would be expected to have higher reproductive success in terms of contribution to future regeneration of the population." If RA is to be meaningful to the evolutionary ecology of plant populations, it should be clearly associated with fitness (Reekie, 1999). Unfortunately, definitions of fitness may be difficult to develop, especially in clonal species where traditional measures based on sexually produced seeds or fruits are inappropriate (Pedersen and Tuomi, 1995; Wikberg, 1995; Winkler and Fischer, 1999).

For the annual *Triplasis purpurea,* a measure of relative fitness based on seed output is reasonable, since the recruitment of new individuals is only possible from a seed. Cheplick and Wickstrom (1999) performed a competition experiment designed to investigate the effect of a large-seeded species, *Cenchrus tribuloides,* on the growth and reproduction of the smaller-seeded *T. purpurea.* Both plants are widely distributed native annual grasses that commonly co-occur along coastal beaches of the Atlantic seashore. Both vegetative mass and RA varied greatly for *T. purpurea* alone or in competition with *C. tribuloides.* The data of Cheplick and Wickstrom (1999) were used to perform linear regressions between RA and vegetative mass, fitness and vegetative mass, and fitness and RA (all log-transformed) for plants in the control and competition treatments. Relative fitness was defined as total seed produced by an individual divided by the total seed produced by the individual showing the highest seed output.

The mean ($\pm$SE) RA of plants in the control (0.16 ± 0.02) did not differ significantly from the mean RA of plants in competition (0.17 ± 0.03, $t = 0.67$, df = 24, $P > 0.05$), even though vegetative mass was significantly greater in the control (651.3 ± 62.2 mg versus 333.01 ± 30.8 mg in competition; $t = 4.5$, df = 24, $P < 0.001$). As is typical for annual plants (see Section VI), RA was not correlated with vegetative mass in the control ($r^2 = 0.09$, $P \gg 0.05$) or in competition ($r^2 < 0.01$, $P \gg 0.05$). However, fitness was significantly correlated with vegetative mass, but only for control plants ($r^2 = 0.51$, $P < 0.01$; fitness = -4.38 [vegetative mass]$^{1.32}$; Fig. 4.3). More importantly, relative fitness was significantly correlated with RA in both the control ($r^2 = 0.72$, $P < 0.001$; fitness = 0.29 [RA]$^{1.19}$) and in competition ($r^2 = 0.78$, $P < 0.001$; fitness = -0.31 [RA]$^{0.81}$; Table 4.1, Fig. 4.3). Clearly, in this annual species greater allocation to reproduction typifies individuals with the highest relative fitness regardless of the competitive environment. In nature, if significant heritability exists for RA, directional selection for increased RA may occur if individuals with high fitness contribute most to the next generation's gene pool.

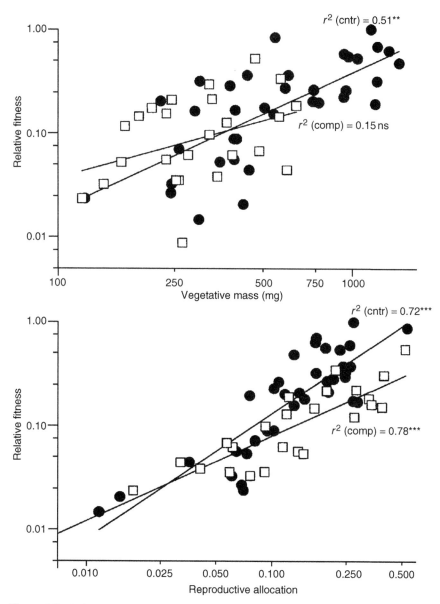

Figure 4.3 Relation of relative fitness to vegetative mass (upper) and reproductive alloca-tion (lower) for *Triplasis purpurea* growing alone (cntr) or in competition (comp) with *Cenchrus tribuloides*. Data collected during the study by Cheplick and Wickstrom (1999) were used to construct the figure. See text and Table 4.1 for regression results.

Table 4.1 Selected Examples of the Relation of Relative Fitness (y) to RA (x), Showing Variation in Slopes and y-intercepts. Separate Regressions are Depicted for each of 10 Replicated Clones and for the Entire Data Set in *Amphibromus scabrivalvis*

Species	Life form	Experimental conditions	n	y-intercept	Slope	r^2	Significance	Source
Triplasis purpurea	Annual grass	Control	36	0.287	1.188	0.717	$P < 0.001$	Cheplick and Wickstrom, 1999
		Competition	25	−0.307	0.806	0.779	$P < 0.001$	
Plantago asiatica	Perennial herb	Altitudinal gradient	12	0.854	2.435	0.690	$P < 0.01$	Kawano and Matsuo, 1983
Amphibromus scabrivalvis	Rhizomatous, perennial grass	Variable soil nutrients	222	0.202	12.530	0.220	$P < 0.01$	Cheplick, 1995
		Clone B	23	0.217	12.277	0.216	$P < 0.05$	
		Clone D	22	0.461	3.812	0.007	Not significant	
		Clone E	24	0.184	14.389	0.306	$P < 0.01$	
		Clone F	24	0.232	14.522	0.182	$P < 0.05$	
		Clone G	21	0.226	8.826	0.183	$P < 0.05$	
		Clone H	19	0.091	16.472	0.310	$P < 0.01$	
		Clone I	22	0.246	7.528	0.088	Not significant	
		Clone J	22	0.121	13.161	0.519	$P < 0.001$	
		Clone K	23	0.093	16.739	0.504	$P < 0.001$	
		Clone M	22	0.056	26.138	0.343	$P < 0.01$	

While the relative fitness of annual plants may be closely coupled to RA, what about perennial species? Detailed studies of the reproductive ecology of the perennial ruderal *Plantago asiatica* provided data on mean seed output, RA, and plant biomass for populations from 10 habitats over an altitudinal gradient in Japan (Kawano and Matsuo, 1983). RA was based on the summed dry masses of flowers, capsules, seeds, and scapes. Linear regression of relative fitness on RA (both log-transformed) yielded a remarkably tight correlation across the diverse habitats ($r^2 = 0.69$, $P < 0.01$, fitness = 0.85 $[RA]^{2.43}$; Table 4.1, Fig. 4.4). Although there were no consistent correlations between biomass and RA for the individual populations, larger plants produced more seeds (Kawano and Matsuo, 1983) and mean relative fitness across habitats was correlated with plant mass ($r^2 = 0.46$, $P < 0.01$, fitness = −0.97 $[mass]^{1.08}$; Fig. 4.4). Clearly, in this weedy colonizer, habitats conducive to a high RA result in a greater seed output relative to less favorable habitats.

A similar attempt was made to relate relative fitness to RA in the rhizomatous perennial *Amphibromus scabrivalvis*, which normally inhabits open grassland communities in South America (Swallen, 1931). As indicated earlier, sexual reproduction via seeds may not be critical to the fitness of clonal plants. In a greenhouse study designed to investigate the plasticity of clonal growth and sexual reproduction, allocation to seeds (based on clone dry mass) was typically low, ranging from 1 to 2.5% for 10 clones (Cheplick, 1995). Seed numbers were recorded per clone so that relative (sexual) fitness was calculated as before. Across all replicates of all clones ($n = 222$), there was a positive relationship of fitness to RA, but the regression only explained 22% of the variance in fitness (Table 4.1). However, the strength of this relationship varied from clone to clone, suggesting genotypic variation in the extent to which relative fitness depended on RA. For some clones (e.g., D and I in Table 4.1), greater allocation to sexual reproduction did not result in higher relative fitness. Perhaps the allocation of resources to asexual modules (ramets) would be more tightly coupled to clone "fitness" in long-lived perennial herbs.

V. Allometry of Modules

Given the modular construction of plants (White, 1979; Vuorisalo and Mutikainen, 1999), and the fact that a module may have its own reproductive structures, it might be surmised that modules would show allometric relationships with RA. The resources available for reproduction by a module are likely to predominantly originate in the module itself. In addition, the allometry of RA for modules of the same type or developmental age may contribute differentially to the allometry of RA for the complete plant. Shipley and Dion (1992) recognized this when they explored the

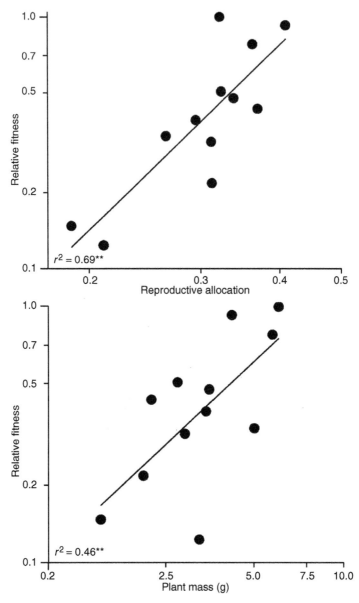

Figure 4.4 Relation of relative fitness to RA (upper) and plant mass (lower) for *Plantago asiatica*. Mature plants at the fruiting stage were collected from habitats along an altitudinal gradient in Japan. Data are from Table 2 of Kawano and Matsuo (1983). See text and Table 4.1 for regression results.

allometry of seed production *per ramet* in 57 herbaceous species and related it to the vegetative mass of ramets rather than to entire genets.

Although clonal plants that show subdivision of each genet into a series of sexually reproducing ramets may be an obvious choice for the study of the allometry of modules, the ramets often have their own root system and may function somewhat independently of other ramets of the same genet (Marshall and Price, 1997). Aclonal, annual species can also be distinctly modular, producing flowers, fruits, and seeds at nodes along an indeterminate stem axis (Cheplick, 2001). However, because an individual has a single root system, one might surmise that modules have the opportunity to receive similar water and mineral resources. Nevertheless, variation can occur along the modules due to developmental and position-dependent effects, and perhaps, micro-environmental heterogeneity in space and time.

The summer annual, invasive grass *Microstegium vimineum* provides an opportunity to examine the reproductive allometry of modules produced in a distinct developmental sequence. The species shows axillary cleistogamy, producing cleistogamous (CL) spikelets along racemes enclosed by the leaf sheath (Fig. 4.5). Each tiller can also produce an emergent raceme of chasmogamous (CH) spikelets at the terminal node (in addition to an axillary CL raceme). Development of racemes along a tiller is basipetal and each node progressively further back from the terminal node contains developmentally younger CL racemes (Fig. 4.5). For example, in autumn when terminal CH and CL racemes contain mature seeds, racemes at subterminal nodes 3 and 4 are likely to contain immature CL spikelets without seeds (Cheplick, unpublished data). Each upper subterminal module along a tiller consists of a culm, leaf, and CL raceme. The terminal module also has these three components plus a CH raceme.

For a study of the developmental and reproductive biology of *M. vimineum*, seeds collected from a secondary forest in central New Jersey, USA, were stratified and then germinated to provide seedlings. Plants were grown in a 1:1 mixture of topsoil and fine vermiculite in plastic pots ($8 \times 8 \times 7.4$ cm) in a temperature-controlled greenhouse during the summer and fall of 2002. Although details of this study will be reported elsewhere, for the present purpose, detailed dry mass data were obtained on a subsample of 20 plants at the time of normal senescence (October). The primary tiller of each plant was carefully sectioned into its component modules (Fig. 4.5). Reproductive allocation *per module* was defined as the mass of the CL or CH raceme divided by the mass of the culm plus leaf. Culm and leaf mass collectively represent the module's vegetative mass. The RA of CL and CH for the complete tiller (all modules combined) was also determined.

For the complete tiller, mean ($\pm$SE) allocation to CL (0.0506 ± 0.0038) was greater than allocation to CH (0.0213 ± 0.0011; $t = 7.51$, df $= 19$, $P < 0.0001$). However, only CL allocation showed a significant decrease with increasing vegetative mass ($r^2 = 0.50$, $P < 0.01$,

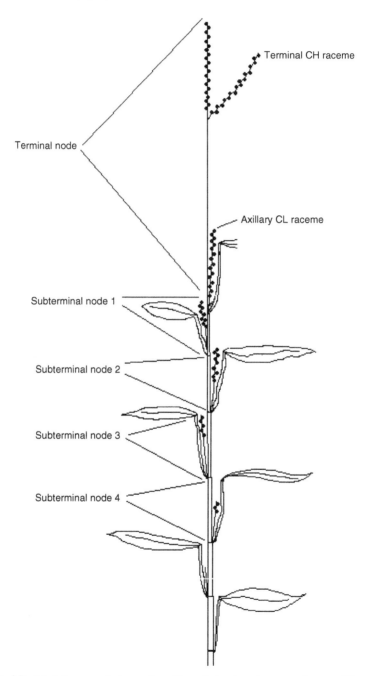

Figure 4.5 Modular morphology of a tiller of the invasive annual grass *Microstegium vimineum*, a species with axillary cleistogamous (CL) spikelets and terminal chasmogamous (CH) spikelets. The lowermost nodes do not produce racemes.

CL = 1.34 [vegetative mass]$^{-0.92}$; Fig. 4.6). Allocation to CH was not correlated with vegetative mass ($r^2 = 0.03$, $P \gg 0.05$) at the whole tiller level.

For the terminal node, allometric relationships for CH and CL were different. Both CH and CL allocation were negatively correlated with the vegetative mass of the module (CH: $r^2 = 0.34$, $P < 0.01$, CH = 0.61 [vegetative mass]$^{-0.63}$; CL: $r^2 = 0.20$, $P < 0.05$, CL = 0.38 [vegetative mass]$^{-0.63}$; Fig. 4.6). At this node, CH allocation (0.4132 ± 0.0348) was significantly greater than CL allocation (0.2642 ± 0.0343; $t = 3.51$, df = 19, $P < 0.01$). Thus, on a whole tiller level, allocation to CH is relatively constant in relation to tiller size, while a trade-off exists for CL allocation because larger tillers allocate proportionately more to CH relative to CL reproduction (Fig. 4.6). At the terminal node, a heavier vegetative module invariably results in lower allocation to both CH and CL.

For the subterminal nodes which have sheath-enclosed CL racemes, allocation to developmentally immature spikelets at nodes 3 and 4 was unrelated to the vegetative mass of modules (node 3: $r^2 = 0.03$, $P \gg 0.05$; node 4: $r^2 < 0.01$, $P \gg 0.05$; Fig. 4.7). For the two nodes immediately below the terminal node, allocation to CL was negatively correlated with module vegetative mass (node 1: $r^2 = 0.37$, $P < 0.01$, RA = 1.33 [vegetative mass]$^{-1.25}$; node 2: $r^2 = 0.20$, $P < 0.05$, RA = 1.06 [vegetative mass]$^{-1.13}$). It is apparent from these analyses of the allometry of modules that the overall negative relationship of CL allocation to the vegetative mass of a tiller was predominantly due to the mature, seed-producing spikelets at the uppermost nodes. In this annual species, the allometric relations of RA to module mass changes with developmental age of individual modules. Future research on the allometry of RA should consider modules, ramets, or plants at similar developmental ages or stages when the objective is to interpret environment-dependent phenotypic variation (Coleman *et al.*, 1994; Acosta *et al.*, 1997).

VI. Allometry of RA and Plant Life History

Annuals: Quite a few investigations have been made into the allometry of reproduction in annual species. While some of these simply correlate reproductive output with plant mass, others examine the relationship of RA to mass. Both increasing (Kawano and Nagai, 1986; Sans and Marsalles, 1994) and decreasing (Samson and Werk, 1986; Cheplick, 1994; Sugiyama and Bazzaz, 1998) RA with vegetative mass have been reported. Vega *et al.* (2000) showed that allometric relationships could be quite complex for three annual crops: RA was lowest in the smallest plants, remained constant over a range of "mid-size" plants, and diminished for the largest plants! However, the most common pattern for annuals may be *no* significant relationship between RA and mass (Harper and Ogden, 1970;

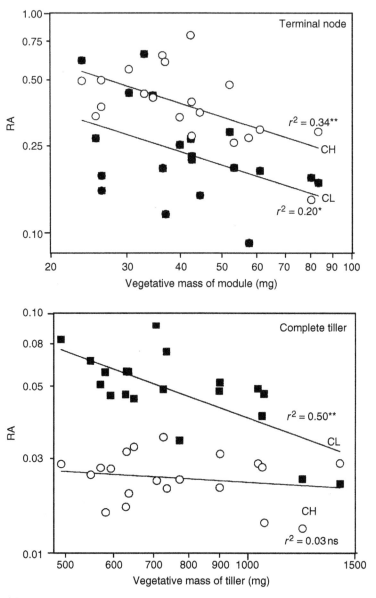

Figure 4.6 Relation of CH and CL reproductive allocation (RA) to the vegetative mass of the terminal module (upper) and to the vegetative mass of the complete tiller (lower) for *Microstegium vimineum.* See text for regression results.

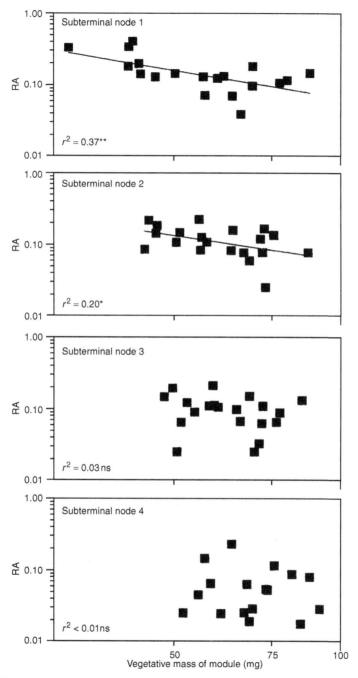

Figure 4.7 Relation of RA to the vegetative mass of subterminal modules (see Fig. 4.5) in *Microstegium vimineum*. See text for regression results.

Kawano and Miyake, 1983; Hara *et al.*, 1988; Kawano *et al.*, 1989; Wagner, 1989; Cheplick, 2001).

The divergent patterns in allometry among annual species are relatively easy to document, but difficult to explain. Values of RA clearly depend on the phenological stages examined (Hara *et al.*, 1988; Sadras *et al.*, 1997). Differences among studies in the inclusion or exclusion of ancillary reproductive structures also contribute to variation in the reported RA–mass allometry. Annuals with opportunistic, indeterminate growth may be able to increase allocation to flowers, fruits, and seeds as larger plants add more reproductive modules over time (Sans and Marsalles, 1994). For other species where a similar mass of reproductive propagules are made at each node along an indeterminate, growing stem, the RA for each module is relatively fixed and the RA–mass relation will be stable (Cheplick, 2001).

The reproductive expenditure, as defined earlier (Section III), is typically lowest for the largest plants. For example, in the competition study of annual dune species described earlier (Cheplick and Wickstrom, 1999), the size of *Triplasis purpurea* plants varied greatly, but mass and RA were not correlated (Table 4.2); however, expenditure was a decreasing function of vegetative mass ($r^2 = 0.74$, $P < 0.001$, expenditure $[\times 10^3] = 2.73$ [vegetative mass]$^{-0.87}$; Fig. 4.8). Across increasing soil nutrient treatments of *Coix ma-yuen* (Kawano *et al.*, 1989), RA increased in large plants, but expenditure declined dramatically (Fig. 4.2). Because it can be shown that expenditure is, by definition, the ratio of individual seed mass to vegetative mass, and plant size typically shows larger variation in relation to environmental heterogeneity than does average seed mass, it is not surprising that expenditure decreases with increasing plant size. Because energy expended on individual propagules declines as plants grow larger, it might be surmised that large plants can allocate more to vegetative tissues. In the weedy annual *Amaranthus albus,* for example, allocation to leaves and roots showed a significant decrease with increasing RA (Cheplick, 2001).

Herbaceous perennials: The allometry of RA has been studied in a fair number of herbaceous perennials, but often for different reasons. For example, one researcher might be interested in the allometry of RA in reference to the cost of reproduction (Worley and Harder, 1996), while another might be concerned with allometric relationships that arise from intraspecific competition (van Kleunen *et al.*, 2001). As would be surmised, the diversity of objectives has produced a plethora of approaches and a myriad of patterns! The allometry of RA with vegetative mass can vary among related species, and can differ among populations within a species (Table 4.2). For two *Plantago* species, RA decreased with vegetative mass in *P. major* (Reekie, 1998), while there were no consistent correlations between RA and mass in multiple populations of *P. asiatica* (Kawano and Matsuo, 1983). In *Poa pratensis,* a rhizomatous perennial grass, RA

Table 4.2 Selected Examples of the Allometry of RA, Showing Variation in Slopes and y-intercepts. Reproductive Allocation (y) was Regressed onto Vegetative mass (x). Data were Log$_{10}$-transformed for *A. scabrivalvis* and *T. purpurea*. Separate Regressions are Depicted for Each of 10 Replicated Clones and for the Entire Data Set in *A. scabrivalvis*

Species	Life form	Experimental conditions or site	n	y-intercept	Slope	r^2	Significance	Source
Helianthus annuus	Annual herb	4 m^{-2}	37	—[a]	—[a]	—[a]	Not significant	Kawano and Nagai, 1986
		11 m^{-2}	54	27.009	7.964	0.354	$P < 0.05$	
		25 m^{-2}	90	11.340	14.353	0.626	$P < 0.001$	
		100 m^{-2}	117	17.418	8.396	0.391	$P < 0.001$	
		400 m^{-2}	53	17.812	12.552	0.708	$P < 0.001$	
Triplasis purpurea	Annual grass	Control	36	-1.997	0.404	0.095	Not significant	Cheplick and Wickstrom, 1999
		Competition	25	-1.207	0.111	0.002	Not significant	
Alliaria petiolata	Biennial herb	Site 1	41	0.592	0.096	0.017	Not significant	Susko and Lovett-Doust, 2000
		Site 2	90	0.450	0.028	0.002	Not significant	
		Site 3	132	0.641	0.014	0.002	Not significant	
Taraxacum officinale	Perennial herb	Agricultural field, undisturbed field	32	0.16	0.13	0.63	$P < 0.001$	Welham and Setter, 1998
			29	0.73	0.02[b]	0.17	$P = 0.025$	
Amphibromus scabrivalvis	Rhizomatous, perennial grass	Variable soil nutrients	222	-1.930	-0.642	0.319	$P < 0.001$	Cheplick, 1995
		Clone B	23	-1.842	-0.604	0.281	$P < 0.01$	
		Clone D	22	-1.943	-0.440	0.330	$P < 0.01$	
		Clone E	24	-1.938	-0.661	0.361	$P < 0.01$	
		Clone F	24	-1.919	-0.513	0.284	$P < 0.01$	
		Clone G	21	-1.786	-0.927	0.552	$P < 0.01$	
		Clone H	19	-1.955	-0.653	0.299	$P < 0.05$	
		Clone I	22	-2.017	-0.732	0.530	$P < 0.01$	
		Clone J	22	-1.835	-0.933	0.476	$P < 0.01$	
		Clone K	23	-1.888	-0.730	0.297	$P < 0.01$	
		Clone M	22	-2.121	-0.763	0.263	$P < 0.05$	

[a] Regression results not provided in the original source.
[b] Regression model also included a 2nd order polynomial term ($-0.005x^2$).

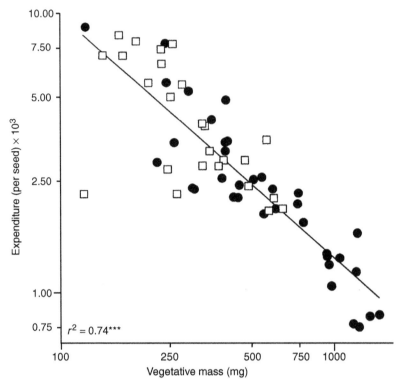

Figure 4.8   Relation of reproductive expenditure per seed to vegetative mass for *Triplasis purpurea* growing alone (filled circles) or in competition with *Cenchrus tribuloides* (open squares). Data were calculated from results of Cheplick and Wickstrom (1999). See text for regression results.

was not related to mass in a population from an unmowed lawn, but was positively correlated with mass in an old field population one year and negatively correlated the second year (Wagner, 1989).

RA did not depend on plant size in the biennial *Alliaria petiolata* (Table 4.2, Susko and Lovett-Doust, 2000) or in the perennial *Ranunculus acris* (Hemborg and Karlsson, 1998). However, a summary of studies that included 10 perennial herbs in Japan (Hara *et al.*, 1988) revealed that RA showed a "conspicuous decreasing trend with increase in individual biomass." RA was also negatively correlated with plant size among 15 perennial grass species (Wilson and Thompson, 1989). In the clonal perennial *Saxifraga hirculus*, RA tended to be greatest for small ramets in four of five habitats in Sweden (Ohlson, 1988). RA also decreased with plant size in *Trollius europaeus* (Hemborg and Karlsson, 1998) and *Rumex obtusifolius* (Pino *et al.*, 2002). In contrast, for four clonal perennial herbs, sexual RA increased with ramet size (Hartnett, 1990), and additional positive relationships between RA and mass have been described for *Arum italicum* (Mendez and Obeso, 1993), *Taraxacum officinale* (Table 4.2, Welham and Setter, 1998), and a few other perennials (Samson and Werk, 1986).

Hartnett (1990) suggested that in nonclonal perennials, RA would decrease as size increased because the costs of mechanical support tissues will increase disproportionately. On a per-ramet basis in clonal species, RA was an increasing function of size, because ramets were unlikely to reach a size "at which they incur disproportionately large costs associated with biomechanical support," whereas genet size was a poor predictor of RA (Hartnett, 1990). Using seed allocation and vegetative mass data for clones of the rhizomatous perennial *Amphibromus scabrivalvis* that varied in size due to fertilizer treatments and variable ages at harvest (Cheplick, 1995), it was found that RA showed a significant decrease with mass (Table 4.2, Fig. 4.9). In this study, variation in size among replicates of each clone should only be due to environmental conditions (i.e., soil nutrients) or age at harvest. Unfortunately, seed output was not recorded per ramet and therefore, it cannot be determined whether per-ramet RA increased or decreased with ramet mass.

Reekie (1998) proposed that higher per gram cost of capsule production for larger plants could explain the observed negative correlation between RA and size in *Plantago major,* a rosette-forming perennial. However, when reproductive expenditure was calculated for the 10 clones of the rhizomatous *A. scabrivalvis* (Cheplick, 1995), it showed a significant decline with increasing vegetative mass in (a) unfertilized replicates at 26 weeks ($r^2 = 0.63$, $P < 0.01$, expenditure $[\times 10^3] = 0.65$ [vegetative mass]$^{-0.72}$), (b) fertilized replicates at 26 weeks ($r^2 = 0.86$, $P < 0.001$, expenditure $[\times 10^3] = 0.57$ [vegetative mass]$^{-0.94}$), and (c) fertilized replicates at 34 weeks ($r^2 = 0.76$, $P < 0.001$, expenditure $[\times 10^3] = 0.51$ [vegetative mass]$^{-0.90}$; Fig. 4.10). Since these analyses involved genet means, they represent genetic correlations. Thus, selection for larger plants would indirectly select for a lower expenditure and RA. In a competitive field environment, it may be advantageous for a clonal species to allocate more resources to vegetative tissues needed for clone maintenance and expansion rather than sexual reproduction, thereby reducing both expenditure and RA.

Woody perennials: Although there is much allometric information on woody plants involving scaling relations between morphological variables such as trunk diameter and tree size (Whittaker and Woodwell, 1968; Niklas, 1994), only a few studies have reported on the allometry of sexual reproduction in detail (Peters *et al.,* 1988; LeMaitre and Midgley, 1991; Acosta *et al.,* 1993, 1997; Suzuki, 2000, 2001). Clearly, in woody species, much of the standing biomass is contained in supporting tissues, and RA is perhaps best expressed on an annual basis as the ratio of reproductive to leaf dry mass. In addition, due to the modular nature of shoot production in woody

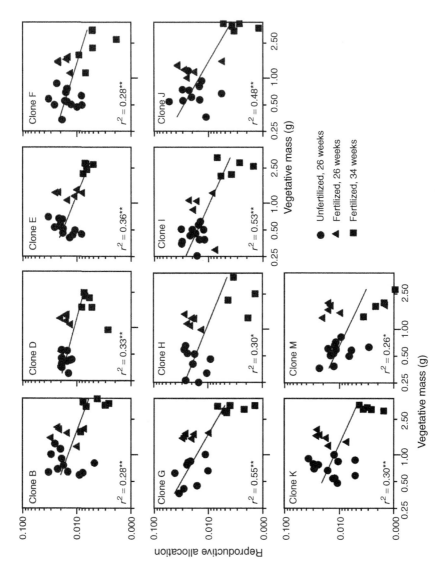

Figure 4.9   Relation of RA to vegetative mass for 10 clones of *Amphibromus scabrivalvis*. Replicates of each clone were grown for 26 weeks in unfertilized or fertilized pots or for 34 weeks in fertilized pots. Data were derived from the study reported in Cheplick (1995). See Table 4.2 for regression results.

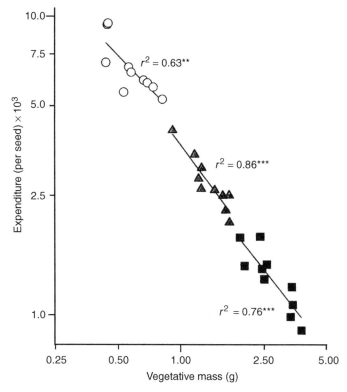

Figure 4.10 Relation of reproductive expenditure to vegetative mass for 10 clones of *Amphibromus scabrivalvis*. Replicates of each clone were grown for 26 weeks in unfertilized (circles) or fertilized pots (triangles) or for 34 weeks in fertilized pots (filled squares). Data were derived from the study reported in Cheplick (1995). See text for regression results.

plants, reproductive output can be quite heterogeneous among hierarchical levels of organization such as branches, shoots, and individuals (Acosta *et al.*, 1993; Vuorisalo and Mutikainen, 1999; Suzuki, 2001).

Peters *et al.* (1988) reported significant allometry of fruit crop mass to stem diameter for 22 species of shrubs and trees. They also reported a decline in the ratio of fruit to shoot mass, but an increase in fruit to leaf mass ratio, with increasing stem diameter. They suggested that the former reflected the "overwhelming influence of nonphotosynthetic tissues" on shoot mass. Hence, larger woody plants may need to invest relatively more in supporting tissues, as reproductive structures become heavier. As with some herbaceous perennials, RA (based on reproductive to leaf mass ratio) showed an increase with increasing plant size. Across 35 species of *Protea* from South Africa, inflorescence mass showed significant allometry with both leaf mass and stem diameter (LeMaitre and Midgley, 1991). The species with the heaviest inflorescence could produce an inflorescence nearly 16 times as large as another species having the same leaf mass because of the thick diameter of the woody stem!

To explore the possible allometry of RA to vegetative (leaf) mass for woody perennials, data on mean annual seed production, seed mass, and leaf mass for canopy trees and shrubs were extracted from Greene and

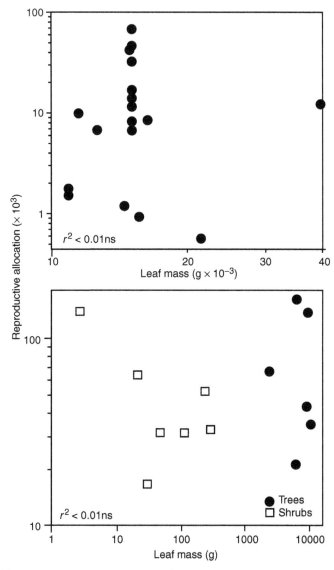

Figure 4.11   Relation of RA to leaf mass of 17 tree species (upper), and for six tree (solid circles) and six shrub species (open squares) (lower). Data were formulated from Tables 1 and 2 of Greene and Johnson (1994).

Johnson (1994). As might be expected, RA values were extremely low relative to per capita leaf mass (Fig. 4.11). Furthermore, for the tree species compiled in Table 1 of Green and Johnson (1994), there was no relationship between RA and leaf mass ($r^2 < 0.01$, $P \gg 0.05$). There was also no allometric relationship detected for the six tree and six shrub species compiled in Table 2 of Greene and Johnson (1994) ($r^2 < 0.01$, $P \gg 0.05$; Fig. 4.11). From this crude analysis, RA could not be allometrically related to the available annual photosynthetic production at the whole-plant level. Future work should better refine the level of analysis by considering the production of flowers, fruits, and seeds for newly produced modules on an annual basis (Niklas and Enquist, 2003).

VII. Determinants of Allometry

Decomposing vegetative mass components: As is evident from the preceding discussion, the choice of what constitutes "vegetative mass" for the denominator of the allocation ratio will certainly impact the allometry of RA. Path analysis provides a superb tool for decomposing the relative effects of a group of inter-related, endogenous variables on RA (Shipley, 2000). A path model specifies a presumed causal structure of the relationships among measured variables, allowing one to separate direct and indirect effects. In the path diagram, a one-headed arrow denotes an effect of one variable on another, while a two-headed arrow indicates a correlation between two variables. Standardized partial regression coefficients (path coefficients) are generated during path analysis (details in Shipley, 2000).

As an example of the utility of this approach, leaf, stem, and root mass of the weedy annual *Amaranthus albus* were employed to ascertain their inter-relationships and relative contribution to RA under low and high soil nutrients in a greenhouse experiment (Cheplick, 2001). As shown in Fig. 4.12, mass components were intercorrelated, but stem mass was most directly related to RA. In fact, under both nutrient treatments, high leaf and root mass resulted in lower RA, as indicated by negative path coefficients (Fig. 4.12). Because the seeds are continually matured along indeterminate branches in *A. albus,* increased RA of larger plants was mostly caused by greater number, length, and mass of stems (Cheplick, 2001). It should be worthwhile to decompose vegetative mass via path analysis in other species in future studies of the allometry of RA.

Indirect effects of other plant traits: Investigations into the allometry of RA should also explore the many other interacting plant traits that typically impact vegetative mass. Not surprisingly, both morphological and physiological traits have been shown to influence vegetative mass and consequently, individual reproductive fitness (Farris and Lechowicz, 1990; Callahan and Waller, 2000; Gibson, 2002). For example, in the annual *Triplasis purpurea,*

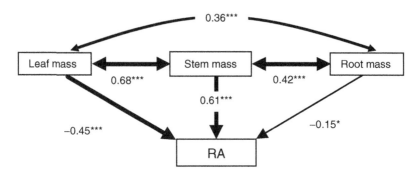

(a) Low soil nutrients

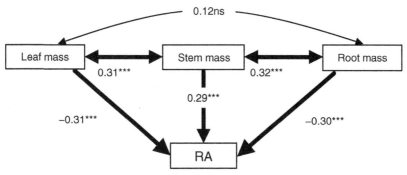

(b) High soil nutrients

Figure 4.12   Decomposition of the vegetative mass components of allometry by path analysis of RA in *Amaranthus albus*. Plants were grown under low (a) or high (b) soil nutrients. Path coefficients are presented along with statistical significance ($*P < 0.05$; $***P < 0.0001$). Cheplick, G. (2001), *Int. J. Plant Sci.* **162**, 807–816, Fig. 4, © 2001 by The University of Chicago. All rights reserved.

early size, life span, and the number of tillers all indirectly impacted the number of seeds produced via their effects on vegetative mass (Cheplick and White, 2002). In all of these studies, path analysis has again proven useful in determining the relative contributions of various plant traits to reproductive success. Future research into the allometry of RA should employ such analyses to critically evaluate the relative contribution of morphological and physiological traits to the evolution of reproductive strategies in plants.

VIII. Conclusions

The principle of resource allocation has been central to the development of life history theory in plants. Patterns in reproductive allocation, in particular, have been described in relation to populations and species, habitats and environmental conditions, and experimental treatments. As should be clear from this review, a substantial fraction of the natural variation observed in RA can apparently be explained by variation in plant size. Where genetically based variation in the size-dependence of reproduction occurs (e.g., Schmid and Weiner, 1993), the potential exists for natural selection to shape allometric relations in an adaptive manner, at least within the constraints set by physiology and development.

In conclusion, eight salient points may be made.

(1) The allometry of RA should be considered an evolved component of a species reproductive strategy.
(2) The evolution of a particular pattern of RA in relation to environmental conditions or life history may reflect prior selection for quantitative traits that mostly impact vegetative size.
(3) For the few species examined to date, relative fitness based on seed output is positively correlated with both RA and vegetative mass, but correlations are strongest for annuals and weaker in clonal perennials.
(4) The allometry of RA depends on developmental age and individual reproductive modules can contribute differentially to the allometric relationship for the complete plant.
(5) Reproductive expenditure, the proportional allocation of vegetative mass to an individual propagule, is typically negatively related to vegetative mass in both annual and perennial plants across a broad range of sizes.
(6) For annuals, there is often no significant relationship between RA and vegetative mass, while in herbaceous perennials the relationship is often (but not always) negative.
(7) For the few woody perennials examined to date, RA shows no clear relation to vegetative size based on available annual photosynthetic production at the whole-plant or modular level.
(8) Indirect effects of other plant traits that impact the primary vegetative mass components (leaf, stem, root) should be considered in future studies of the allometry of RA in diverse plant species.

References

Acosta F. J., Delgado J. A., Lopez F. and Serrano J. M. (1997) Functional features and ontogenetic changes in reproductive allocation and partitioning strategies of plant modules. *Plant Ecol* **132**: 71–76.

Acosta F. J., Serrano J. M., Pastor C. and Lopez F. (1993) Significant potential levels of hierarchical phenotypic selection in a woody perennial plant, *Cistus ladanifer*. *Oikos* **68**: 267–272.

Armstrong R. A. (1982) A quantitative theory of reproductive effort in rhizomatous perennial plants. *Ecology* **63**: 679–686.

Bazzaz F. A. (1996) *Plants in Changing Environments: Linking Physiological, Population, and Community Ecology.* Cambridge University Press, Cambridge.

Bazzaz F. A. (1997) Allocation of resources in plants: state of the science and critical questions. In *Plant Resource Allocation* (F. A. Bazzaz and J. Grace, eds.) pp. 1–37, Academic Press, San Diego.

Bazzaz F. A. and Ackerly D. D. (1992) Reproductive allocation and reproductive effort in plants. In *Seeds: The Ecology of Regeneration in Plant Communities* (M. Fenner, ed.) pp. 1–26, CAB International, Wallingford.

Callahan H. S. and Waller D. M. (2000) Phenotypic integration and the plasticity of integration in an amphicarpic annual. *Int J Plant Sci* **161**: 89–98.

Cheplick G. P. (1994) Life history evolution in amphicarpic plants. *Plant Species Biol* **9**: 119–131.

Cheplick G. P. (1995) Plasticity of seed number, mass, and allocation in clones of the perennial grass *Amphibromus scabrivalvis. Int J Plant Sci* **156**: 522–529.

Cheplick G. P. (2001) Quantitative genetics of mass allocation and the allometry of reproduction in *Amaranthus albus*: relation to soil nutrients. *Int J Plant Sci* **162**: 807–816.

Cheplick G. P. and White T. P. (2002) Saltwater spray as an agent of natural selection: no evidence of local adaptation within a coastal population of *Triplasis purpurea* (Poaceae). *Amer J Bot* **89**: 623–631.

Cheplick G. P. and Wickstrom V. M. (1999) Assessing the potential for competition on a coastal beach and the significance of variable seed mass in *Triplasis purpurea. J Torrey Bot Soc* **126**: 296–306.

Coleman J. S., McConnaughay K. D. M. and Ackerly D. D. (1994) Interpreting phenotypic variation in plants. *Trends Ecol Evol* **9**: 187–191.

Damuth J. (2001) Scaling of growth: plants and animals are not so different. *PNAS* **98**: 2113–2114.

Farrar J. and Gunn S. (1998) Allocation: allometry acclimation—and alchemy? In *Inherent Variation in Plant Growth. Physiological Mechanisms and Ecological Consequences* (H. Lambers, H. Poorter, and M.M.I. Van Vuuren, eds.) pp. 183–198, Backhuys Publishers, Leiden.

Farris M. A. and Lechowicz M. J. (1990) Functional interactions among traits that determine reproductive success in a native annual plant. *Ecology* **71**: 548–557.

Gayon J. (2000) History of the concept of allometry. *Amer Zool* **40**: 748–758.

Gibson D. J. (2002) *Methods in Comparative Plant Population Ecology.* Oxford University Press, Oxford.

Gould S. J. (1966) Allometry and size in ontogeny and phylogeny. *Biol Rev* **41**: 587–640.

Gould S. J. (1975) Allometry in primates with emphasis on scaling and the evolution of the brain. *Contri Primatol* **5**: 244–292.

Greene D. F. and Johnson E. A. (1994) Estimating the mean annual seed production of trees. *Ecology* **75**: 642–647.

Hancock J. F. and Pritts M. P. (1987) Does reproductive effort vary across different life forms and seral environments? A review of the literature. *Bull Torrey Bot Club* **114**: 53–59.

Hara T., Kawano S. and Nagai Y. (1988) Optimal reproductive strategy of plants, with special reference to the modes of reproductive resource allocation. *Plant Species Biol* **3**: 43–59.

Harper J. L. and Ogden J. (1970) The reproductive strategy of higher plants. I. The concept of strategy with special reference to *Senecio vulgaris* L. *J Ecol* **58**: 681–698.

Hartnett D. C. (1990) Size-dependent allocation to sexual and vegetative reproduction in four clonal composites. *Oecologia* **84**: 254–259.

Hemborg A. M. and Karlsson P. S. (1998) Altitudinal variation in size effects on plant reproductive effort and somatic costs of reproduction. *Ecoscience* **5**: 517–525.

Huxley J. S. (1932) *Problems of Relative Growth.* MacVeagh, New York.

Kawano S. (1981) Trade-off relationships between some reproductive characteristics in plants with special reference to life history strategy. *Bot Mag Tokyo* **94**: 285–294.

Kawano S. and Matsuo K. (1983) Studies on the life history of the genus *Plantago*. I. Reproductive energy allocation and propagule output in wild populations of a ruderal species, *Plantago asiatica* L., extending over a broad altitudinal gradient. *J Coll Lib Arts, Toyama Univ (Nat Sci)* **16**: 85–112.

Kawano S. and Miyake S. (1983) The productive and reproductive biology of flowering plants. X. Reproductive energy allocation and propagule output of five congeners of the genus *Setaria* (Gramineae). *Oecologia* **57**: 6–13.

Kawano S. and Nagai Y. (1986) Regulatory mechanisms of reproductive effort in plants. I. Plasticity in reproductive energy allocation and propagule output of *Helianthus annuus* L. (Compositae) cultivated at varying densities and nitrogen levels. *Plant Species Biol* **1**: 1–18.

Kawano S., Hayashi S., Arai H., Yamamoto M., Takasu H. and Oritani T. (1989) Regulatory mechanisms of reproductive effort in plants. III. Plasticity in reproductive energy allocation and propagule output of two grass species, *Oryza sativa* cv. *Akihikari* and *Coix ma-yuen* cultivated at varying densities and nitrogen levels, and the evolutionary-ecological implications. *Plant Species Biol* **4**: 75–99.

Klinkhamer P. G. L., Meelis E., de Jong T. J. and Weiner J. (1992) On the analysis of size-dependent reproductive output in plants. *Funct Ecol* **6**: 308–316.

LeMaitre D. C. and Midgley J. J. (1991) Allometric relationships between leaf and inflorescence mass in the genus *Protea* (Proteaceae): an analysis of the exceptions to the rule. *Funct Ecol* **5**: 476–484.

Loehle C. (1987) Partitioning of reproductive effort in clonal plants: a benefit-cost model. *Oikos* **49**: 199–208.

Marshall C. and Price E. A. C. (1997) Sectoriality and its implications for physiological integration. In *The Ecology and Evolution of Clonal Plants* (H. de Kroon and J. van Groenendael, eds.) pp. 79–107, Backhuys Publishers, Leiden.

Mendez M. and Obeso J. R. (1993) Size-dependent reproductive and vegetative allocation in *Arum italicum* (Araceae) *Can J Bot* **71**: 309–314.

Niklas K. J. (1994) *Plant Allometry: The Scaling of Form and Process.* University of Chicago Press, Chicago.

Niklas K. J. and Enquist B. J. (2002a) Canonical rules for plant organ biomass partitioning and annual allocation. *Amer J Bot* **89**: 812–819.

Niklas K. J. and Enquist B. J. (2002b) On the vegetative biomass partitioning of seed plant leaves, stems, and roots. *Am Nat* **159**: 482–497.

Niklas K. J. and Enquist B. J. (2003) An allometric model for seed plant reproduction. *Evol Ecol Res* **5**: 79–88.

Obeso J. R. (2002) The costs of reproduction in plants. *New Phytol* **155**: 321–348.

Ohlson M. (1988) Size-dependent reproductive effort in three populations of *Saxifraga hirculus* in Sweden. *J Ecol* **76**: 1007–1016.

Pedersen B. and Tuomi J. (1995) Hierarchical selection and fitness in modular and clonal organisms. *Oikos* **73**: 167–180.

Peters R. H. (1983) *The Ecological Implications of Body Size.* Cambridge University Press, Cambridge.

Peters R. H., Cloutier S., Dube D., Evans A., Hastings P., Kaiser H., Kohn D. and Sarwer-Foner B. (1988) The allometry of the weight of fruit on trees and shrubs in Barbados. *Oecologia* **74**: 612–616.

Pino J., Sans F. X. and Masalles R. M. (2002) Size-dependent reproductive pattern and short-term reproductive cost in *Rumex obtusifolius*. *Acta Oecologica* **23**: 321–328.

Reekie E. G. (1998) An explanation for size-dependent reproductive allocation in *Plantago major*. *Can J Bot* **76**: 43–50.

Reekie E. G. (1999) Resource allocation, trade-offs, and reproductive effort in plants. In *Life History Evolution in Plants* (T. O. Vuorisalo and P. K. Mutikainen, eds.) pp. 173–193, Kluwer Academic Publishers, Dordrecht.

Reekie E. G. and Bazzaz F. A. (1987) Reproductive effort in plants. 3. Effect of reproduction on vegetative activity. *Am Nat* **129**: 907–919.

Reekie E. G. and Bazzaz F. A. (1992) Cost of reproduction as reduced growth in genotypes of two congeneric species with contrasting life histories. *Oecologia* **90**: 21–26.

Reiss M. J. (1989) *The Allometry of Growth and Reproduction.* Cambridge University Press, Cambridge.

Sadras V. O., Bange M. P. and Milroy S. P. (1997) Reproductive allocation of cotton in response to plant and environmental factors. *Ann Bot* **80**: 75–81.

Samson D. A. and Werk K. S. (1986) Size-dependent effects in the analysis of reproductive effort in plants. *Am Nat* **127**: 667–680.

Sans F. X. and Masalles R. M. (1994) Life-history variation in the annual arable weed *Diplotaxis erucoides* (Cruciferae) *Can J Bot* **72**: 10–19.

Schlichting C. D. and Pigliucci M. (1998) *Phenotypic Evolution: A Reaction Norm Perspective.* Sinauer Associates, Inc., Sunderland, Massachusetts.

Schmid B. and Weiner J. (1993) Plastic relationships between reproductive and vegetative mass in *Solidago altissima. Evolution* **47**: 61–74.

Shipley B. (2000) *Cause and Correlation in Biology.* Cambridge University Press, Cambridge.

Shipley B. and Dion J. (1992) The allometry of seed production in herbaceous angiosperms. *Am Nat* **139**: 467–483.

Soule J. D. and Werner P. A. (1981) Patterns of resource allocation in plants, with special reference to *Potentilla recta* L. *Bull Torrey Bot Club* **108**: 311–319.

Stewart A. J. A., John E. A. and Hutchings M. J. (2000) The world is heterogeneous: ecological consequences of living in a patchy environment. In *The Ecological Consequences of Environmental Heterogeneity* (M. J. Hutchings, E. A. John and A. J. A. Stewart, eds.) pp. 1–8, Blackwell Science, Oxford.

Sugiyama S. and Bazzaz F. A. (1998) Size dependence of reproductive allocation: the influence of resource availability, competition and genetic identity. *Funct Ecol* **12**: 280–288.

Susko D. J. and Lovett-Doust L. (2000) Plant size and fruit-position effects on reproductive allocation in *Alliaria petiolata* (Brassicaceae) *Can J Bot* **78**: 1398–1407.

Suzuki A. (2000) Patterns of vegetative growth and reproduction in relation to branch orders: the plant as a spatially structured population. *Trees* **14**: 329–333.

Suzuki A. (2001) Resource allocation to vegetative growth and reproduction at shoot level in *Eurya japonica* (Theaceae): a hierarchical investment? *New Phytol* **152**: 307–312.

Swallen J. R. (1931) The grass genus *Amphibromus. Amer J Bot* **18**: 411–415.

Thompson D. W. (1917) *On Growth and Form.* Cambridge University Press, Cambridge.

Tseng M. and Rowe L. (1999) Sexual dimorphism and allometry in the giant water strider *Gigantometra gigas. Can J Zool* **77**: 923–929.

Van Kleunen M., Fischer M. and Schmid B. (2001) Effects of intraspecific competition on size variation and reproductive allocation in a clonal plant. *Oikos* **94**: 515–524.

Vega C. R. C., Sadras V. O., Andrade F. H. and Uhart S. A. (2000) Reproductive allometry in soybean, maize and sunflower. *Ann Bot* **85**: 461–468.

Vuorisalo T. and Mutikainen P. (1999) Modularity and plant life histories. In *Life History Evolution in Plants* (T. O. Vuorisalo and P. K. Mutikainen, eds.) pp. 1–25, Kluwer Academic Publishers, Dordrecht, The Netherlands.

Wagner L. K. (1989) Size dependent reproduction in *Poa annua* and *P. pratensis.* In *The Evolutionary Ecology of Plants* (J. H. Bock and Y. B. Linhart, eds.) pp. 273–284, Westview Press, Boulder, Colorado.

Weiner J. (1988a) The influence of competition on plant reproduction. In *Plant Reproductive Ecology: Patterns and Strategies* (J. Lovett Doust and L. Lovett Doust, eds.) pp. 228–245, Oxford University Press, Oxford.

Weiner J. (1988b) Variation in the performance of individuals in plant populations. In *Plant Population Ecology* (A. J. Davy, M. J. Hutchings and A. R. Watkinson, eds.) pp. 59–81, Blackwell Scientific, Oxford.

Welham C. V. J. and Setter R. A. (1998) Comparison of size-dependent reproductive effort in two dandelion (*Taraxacum officinale*) populations. *Can J Bot* **76**: 166–173.

Whittaker R. H. and Woodwell G. M. (1968) Dimension and reproduction relations of trees and shrubs in the Brookhaven Forest, New York. *J Ecol* **56**: 1–25.

White J. (1979) The plant as a metapopulation. *Ann Rev Ecol Syst* **10**: 109–145.

White J. (1981) The allometric interpretation of the self-thinning rule. *J Theor Biol* **89**: 475–500.

White J. and Harper J. L. (1970) Correlated changes in plant size and number in plant populations. *J Ecol* **58**: 467–485.

Wikberg S. (1995) Fitness in clonal plants. *Oikos* **72**: 293–297.

Wilson A. M. and Thompson K. (1989) A comparative study of reproductive allocation in 40 British grasses. *Funct Ecol* **3**: 297–302.

Winkler E. and Fischer M. (1999) Two fitness measures for clonal plants and the importance of spatial aspects. *Plant Ecol* **141**: 191–199.

Worley A. C. and Harder L. D. (1996) Size-dependent resource allocation and costs of reproduction in *Pinguicula vulgaris* (Lentibulariaceae) *J Ecol* **84**: 195–206.

Zhang D.-Y. and Jiang X.-H. (2002) Size-dependent resource allocation and sex allocation in herbaceous perennial plants. *J Evol Biol* **15**: 74–83.

5

Sex-specific Physiology and its Implications for the Cost of Reproduction

Andrea L. Case, Tia-Lynn Ashman

I. Introduction

Female and male sex functions incur different costs, and the balance of investment between them is an important component of life history evolution (Charnov, 1982). In plants, relative reproductive allocation often covaries with ecological context (Obeso, 2002) or other life history traits (Delph, 1999; Shykoff *et al.,* 2003). But measuring sex-specific costs of reproduction in plants is complicated by the fact that almost all are hermaphroditic, investing in both female and male sex functions during their lifetime. Hermaphroditism makes it inherently difficult to isolate the consequences of female versus male investment to future reproduction and survival, even where the sex functions are temporally or spatially segregated within individuals.

Many plant biologists have turned to dioecious species to determine sex-specific costs of reproduction, exploiting the fact that the sex functions are housed in separate individuals to estimate both short-term (somatic) costs and longer-term (demographic) consequences of investment into female versus male function (reviewed in Geber *et al.,* 1999; Obeso, 2002). Such studies offer unparalleled insight, because they are essentially independent contrasts of plants that differ in little more than traits that affect, or are affected by, their reproductive mode. Females and males within species have similar morphological, ecological, and genetic backgrounds, but differ dramatically in the amount and type of resources used for reproduction. They are well suited to address how differences in sex allocation translate into direct costs of reproduction, and how these, in turn, translate into demographic costs. Females invariably incur a greater total cost of reproduction (see Table 1 in Delph, 1999 and Table 3 in Shykoff *et al.,* 2003). The question is how they manage to pay it.

Resource acquisition is an inherently physiological process, and is fundamental to the issues of cost, investment, and allocation. Despite its potential importance, we have a vague understanding of the evolutionary dynamics and adaptive significance of variation in physiological traits (Ackerly *et al.,* 2000; Arntz and Delph, 2001). Comparisons of separate sexes are especially useful for investigating the role of physiology in mitigating the costs of reproduction. But we understand even less about how and why separate sexes should differ physiologically, primarily because what little data exists appear to be species-specific and highly context-dependent (reviewed in Section VI, Dawson and Geber, 1999; Obeso, 2002). Studies of the evolution of ecophysiological traits in general are fraught with difficulty. Because measurements are time consuming and expensive to obtain, sampling is often too limited in scope to provide adequate statistical power for estimates of genetic variation, heritability, etc. (Ackerly *et al.,* 2000). Researchers commonly rely on instantaneous measures of physiological functioning, often on a small amount of leaf tissue and at only a single point in time per individual. However, the highly plastic nature of physiological traits makes it difficult to integrate and generalize patterns detected in instantaneous, leaf-level data. Physiological data are particularly limited for gender dimorphic species, likely because dioecy is phylogenetically associated with a set of life history traits (e.g., large size, woodiness, perenniality; Renner and Ricklefs, 1995), that further challenge physiological measurement, integration, and generalization.

Two recent reviews discuss sexual dimorphism in physiology (Dawson and Geber, 1999; Obeso, 2002). Both focus on fully dioecious species. We argue that it may be as informative to study these characters in plants with newly arisen gender dimorphism (e.g., gynodioecy), where the potential for evolutionary divergence between the sex morphs is somewhat reduced (Poot, 1997; Schultz, 2003; Caruso *et al.,* 2003; A.L. Case & T.-L. Ashman, unpublished data). Unlike dioecious species, females and hermaphrodites of gynodioecious species are more similar in their reproductive costs – both sexes produce fruits and seeds, while only hermaphrodites produce pollen. For example, identifying the short- and long-term costs of producing pollen should be relatively more straightforward when comparing females to hermaphrodites as opposed to males.

In this chapter, we discuss the implications of sex-specific physiology to the cost of sexual reproduction in plants. We begin by describing various dimorphic sexual systems and why they might be useful for linking physiological process to reproductive function, particularly intermediate sexual systems such as gynodioecy. We describe the differential costs of pollen versus seed production in Section III, as well as common costs of flower production and demographic consequences. In Section IV, we describe physiological mechanisms that may help mitigate the costs of reproduction, followed by predicted patterns of sexual dimorphism in physiology based on differential investment (Section V). Section VI describes the potential causes of sex differences in physiology, followed by a final section summarizing the available data on the presence of sex-specific physiology and its relationship to reproductive investment in a variety of sexual systems.

II. Sexual Polymorphisms

Nearly all of the available data on physiological sex differences comes from dioecious species (see Dawson and Geber, 1999, for review). Other sexual systems are likely to provide important data on this issue. First, hermaphroditic systems that have nonoverlapping sex functions may exhibit physiological patterns corresponding to each sex phase. In monoecious species, plants produce varying proportions of unisexual flowers, such that plant sexual phenotype can be either female (pistillate flowers only), male (staminate flowers only), or hermaphrodite (pistillate and staminate flowers). Pistillate and staminate flowers are often spatially segregated in monoecious species (either within or among inflorescences), such that regular patterns of anthesis can result in completely female or male phases within a season. A similar pattern results for sequentially hermaphroditic (diphasic) species, which reproduce as males in one season and females in another.

Intermediate stages of the evolutionary pathway from hermaphroditism to dioecy could provide important clues about when and how sex-specific physiological traits originate. Dioecy is phylogenetically associated with both monoecy and gynodioecy (Weiblen *et al.*, 2000). In gynodioecious species, plants are morphologically either hermaphrodite (perfect-flowered or monoecious) or female (pistillate-flowered). Hermaphrodites of gynodioecious or monoecious species often vary dramatically in their fruit-setting ability (e.g., Ashman, 1999). Those that consistently fail to produce fruits may be better described as functional males, and may possess physiological traits distinct from both functional hermaphrodites and females. Because sex functions entail different costs (see Section III), we expect that the cost of reproduction will vary among the sexual morphs in relation to their relative investment in male versus female function.

III. Costs of Reproduction

The most common approach to measuring cost is to generate a static estimate of biomass or, less often, nutrients invested in reproduction. While these static measures may provide a gross idea of reproductive cost, there can be an uncoupling between these and demographic measures of reproductive cost. When demographic costs are greater than biomass estimates suggest, it can indicate that biomass inadequately reflects cost, that some costs have been overlooked, or that there is a nonlinear relationship between static estimates and true demographic costs. On the other hand, when demographic costs are not detected or are lower than static estimates suggest (e.g., Ramsey, 1997; Ehrlen and van Groenendael, 2001; Jackson and Dewald, 1994; reviewed in Obeso, 2002), there may be compensatory physiological mechanisms that have not been taken into account.

A. Male Costs

The costs of pollen and fertile stamen production are unique to plants with male function. While pollen accounts for only small fraction of plant reproductive biomass in most taxa, it is protein-rich (Roulston *et al.*, 2000), and thus is more concentrated in nutrients (e.g., N and P) than either flowers or seeds (Ashman, 1994a). As a consequence, biomass does not predict nutrient investment in pollen well, and the relative contribution of pollen to reproductive costs varies dramatically with the currency used (e.g., Ashman, 1994a; Carroll and Delph, 1996; Hemborg and Karlsson, 1999). For example, pollen accounts for less than 10% of the carbon but 25% of the nitrogen allocated to reproduction in monoecious *Pinus meziesii* (McDowell *et al.*, 2000), only 4% of the biomass but 7.3% of the phosphorus allocated to reproduction in hermaphrodites of *Sidalcea oregana* (Ashman, 1994a), and ~2% of total plant biomass of male *Silene dioica* plants but up to 8% of their whole plant nutrient pool (Hemborg and Karlsson, 1999). Thus, biomass can grossly underestimate the cost of reproduction through male function, especially if nutrients rather than photosynthates limit physiological functioning. Mineral nutrient-based estimates may more faithfully reflect true costs of male function. For instance, Ashman (1994b) studied a perennial plant whose growth is nutrient-limited *(Sidalcea oregana),* and found that nitrogen and phosphorus investment in current reproduction explained a significant amount of the variation in future reproduction, while biomass investment did not.

Plants that produce pollen may incur two additional costs. First, individuals that have male function (males or hermaphrodites) may suffer greater "opportunity costs" than females of the species, because they invest in pollen (and possibly more in flowers, see Section III.B) early in the flowering season (Delph, 1990). This early investment of essential nutrients in pollen diverts them away from investment in photosynthetic machinery that might otherwise contribute to increased growth (Eckhart and Seger, 1999). As the "principle of compound interest" applies to plant growth, depression of early growth leads to less resource income for subsequent growth. Supporting this view, Eckhart and Chapin (1997) found that the opportunity costs of male function were more pronounced under nutrient-limited conditions. Second, pollen-bearing plants also suffer greater herbivory during flowering than females of the species (Boecklen and Hoffman, 1993; Ågren *et al.*, 1999;

Ashman, 2002). Greater predation can have two consequences for reproductive costs – the direct loss of vegetative and/or reproductive tissue, and
the additional costs associated with compensation for lost organs –
exacerbating the opportunity costs described earlier. Moreover, because
predatory risks increase with increasing investment in pollen and flowers
(Ashman, 2002; Ashman *et al.*, 2004), demographic costs may escalate nonlin-
early with investment in male function in the presence of herbivores.

B. Female Costs

The costs of ovary, seed, and fruit production are specific to plants with female
function. Seeds are rich in carbohydrates, nutrients, and fats (Jordano, 1992;
Ashman, 1994a), and fruits increase in size dramatically during their maturation. The duration of investment into female function is considerably
longer than that in male function because only fruit production extends
beyond flowering. The greater apparency and high nutritional content
of seeds/fruits obliges females to invest in defense against frugivores
and granivores, further increasing their total reproductive costs. Indeed,
reproductive structures often have a greater concentration of defensive
compounds than leaves (e.g., Zangerl and Berenbaum, 1990). Plants with
female function also must invest in structures that support the weight
of developing fruit (e.g., thick inflorescence stalks), as well as those that
aid in the dispersal of fruits or seeds, including fleshy, armored or
ornate pericarps. These additional investments and greater duration of
investment contributes to a greater total cost of reproduction of female
compared to male function. There are at least two physiological mechanisms by which plants may offset the costs of female function, mechanisms
that are not equivalently available to offset the costs of male function (see
Section IV).

C. Common Flowering Costs

Some costs associated with reproduction are common to plants with male
and/or female function. These include investment in flower and inflorescence construction and maintenance, and the production of pollinator
attractants (e.g., petals or volatiles) and rewards (e.g., nectar or resins).
Flower respiratory and nectar costs can be substantial (Pyke, 1991; Ashman
and Schoen, 1994, 1997) and often scale with flower size (e.g., Ashman and
Stanton, 1991; Galen *et al.*, 1999). In both temperate dioecious and gynodioecious species, flowers of males or hermaphrodites often are larger than
those of females (reviewed in Eckhart, 1999), suggesting that floral attraction costs may be greater for plants with male function. However, pistillate
flowers have longer lifetimes than flowers with male function in both dioecious (reviewed in Primack, 1985) and gynodioecious species (Ashman
and Stanton, 1991; Pettersson, 1992), suggesting that maintenance and
reward costs may be equivalent, or even greater in females. A complete

accounting of both dynamic and static costs is required to determine how these factors contribute to total floral costs and whether sexual dimorphism exists.

D. Demographic Costs

Since fruit-producing plants generally make greater reproductive investments than nonfruiting plants (reviewed in Obeso, 2002), they are expected to pay greater demographic costs. These demographic costs may be manifested as lower survival and/or frequency of flowering or slower vegetative growth. Delph (1999) and Obeso (2002) have independently reviewed the data accumulated to date and test this prediction. Delph (1999) summarized data from 32 dioecious species, where females invested more in reproduction than males and the strongest patterns revealed were that females were larger/older at first reproduction (12 out of 19 and 14 out of 16 species, respectively), and grew less than males (18 out of 26 species). The patterns, however, were less clear cut for flowering frequency and longevity. Females were just as likely to flower at the same frequency as males (7 species) as they were to flower less frequently than males (7 species). Similarly, females were as likely to live as long or longer than males (10 species) as they were to die earlier than males (9 species). Obeso (2002) reviewed 103 dimorphic species (91 of which were dioecious) and found that almost all studies showed some sort of demographic dimorphism, which he defined as including total investment in or frequency of reproduction, vegetative propagation, or survival. Across all the studies he surveyed (NB: he did not limit his studies to those showing greater reproduction by females), he found strong evidence for dimorphism in flowering frequency and survival: males reproduced more frequently (28 out of 31 species) and survived longer (13 out of 17 species) than females. He also found that in species with vegetative dimorphism (plant size or relative growth rate), the direction of dimorphism depended on life form – males of woody species were larger and/or grew faster than females, but the reverse was true for herbaceous species.

   In gynodioecious species, one might expect hermaphrodites to bear the greater reproductive cost as they must produce flowers, fruits/seeds, and pollen, whereas females invest only in flowers and fruits/seeds. A quantitative meta-analysis of 28 studies on a total of 14 gynodioecious species shows that the sex morphs produce similar numbers of flowers per plant, but females produce significantly more seeds per plant than hermaphrodites (Fig. 5.1). These results mirror an independent meta-analysis by Shykoff *et al.* (2003), based on a partially overlapping data set comprised of 54 gynodioecious species in 23 plant families. Indeed, this fitness advantage is a predicted requirement for the maintenance of females in gynodioecious populations (reviewed in Charlesworth, 1999). On the other hand, hermaphrodites produce larger flowers than females (reviewed in Eckhart, 1999;

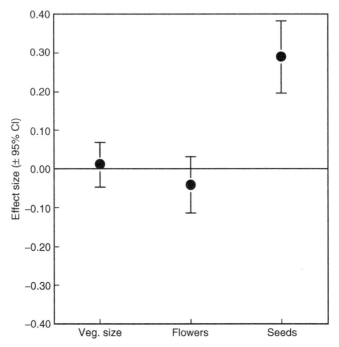

Figure 5.1 Results of meta-analysis of sex-morph differences in vegetative size and repro-duction (flowers per plant, seeds per plant). Data set included 28 studies of 14 gynodioecious species. Effect size (standardized relative sex differences) ± 95% confidence intervals are shown. Traits for which the 95% confidence intervals do not overlap 0 are significant. Studies used in the calculation of these data are listed in the Appendix.

Shykoff *et al.*, 2003). It is difficult to evaluate whether these sex-specific reproductive costs might translate into demographic ones in gynodioecious species, because only a handful of studies have investigated other aspects of gynodioecious life histories. The available data on vegetative size indicates that the sex morphs, on average, attain comparable sizes (Fig. 5.1). In addi-tion, two studies have evaluated vegetative propagation: one study found higher vegetative propagation in hermaphrodites than females (*Fragaria virginiana*, T.-L. Ashman, unpublished data), but the other found no sex difference (*Saxifraga granulata*, Stevens, 1988). Clearly, many more studies are needed to determine if demographic costs in gynodioecious plants follow the pattern predicted from direct costs. It is worth noting, however, that because the presence of females fundamentally changes selection on hermaphrodites, their life history traits may already be altered from their original state, i.e., prior to the evolution of females (see Webb, 1999).

The sexes have likely been selected for different life histories as a result of different reproductive investments (Delph, 1999). However, comparative surveys of demographic differences between the sex morphs can be confounded by historical or environmental factors and cannot distinguish between possible causes for dimorphism in demography. If one wishes to separate a direct demographic cost of reproduction from an evolved response to selection to mitigate the cost of reproduction, one needs to take an experimental approach. For instance, manipulation of reproductive investment via pollination or flower removal (e.g., Ehrlen and Van Groenendael, 2001), or artificial selection (e.g., Delph *et al.,* 2004) can be used to evaluate the direct relationship between reproductive investment and demography. Among-population variation in many gynodioecious species reflects a continuum from newly established gynodioecy to near dioecy (subdioecy) – variation that can be exploited to better understand the relationship between direct and demographic costs of reproduction.

IV. Avenues for Mitigating the Cost of Reproduction

As natural selection favors traits that enhance lifetime reproductive success, any mechanism that reduces cost while maximizing benefit during a reproductive episode should be favored regardless of sexual phenotype. Several nonphysiological mechanisms for minimizing reproductive costs have been previously addressed (and see Delph, 1999); physiological mechanisms should center around the three most important resources common to both reproduction and physiological functioning – carbon, nitrogen, and water. Three physiological avenues may help mitigate the costs of reproduction: (1) produce photosynthetically competent reproductive organs that contribute to their own cost, (2) respond to greater demand by increasing ("up-regulating") vegetative photosynthetic rates, enhancing resource uptake, or increase resource use efficiency, and (3) recoup some of the costs of reproduction through nutrient reabsorption. These avenues for mitigation, however, are not all equally available to male and female function, and, as Fig. 5.2 illustrates, not at all independent from each other.

A. Photosynthetic Reproductive Organs

Ovaries, calyces, and fruits can photosynthesize, although not at the same rates as leaves (e.g., Bazzaz *et al.,* 1979; Galen, 1993). Reproductive structures may not "pay" for themselves entirely, but they can contribute up to 60% of their carbon cost (e.g., Bazzaz *et al.,* 1979; Jurik, 1985; Galen *et al.,* 1993; Hogan *et al.,* 1998; McDowell *et al.,* 2000). Because female sex organs are capable of offsetting their respiratory costs, but male organs are not, reproduction through male function may be as costly as that through female function (Galen *et al.,* 1999, Hogan *et al.,* 1998). Laporte and Delph (1996) found that staminate flowers on males of dioecious *Silene latifolia* (Caryophyllaceae) contribute to their carbon costs only when in bud,

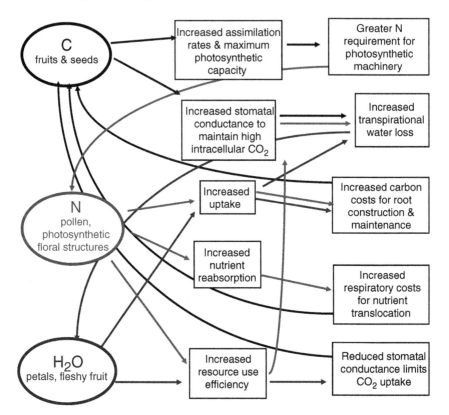

Figure 5.2 Some examples of physiological mechanisms for accommodating proximal costs of reproductive investment by plants. Straight arrows originate from one of three types of resource in greatest demand (left column): C, black = carbon; N, red = nitrogen or other mineral nutrients; H_2O, blue = water. The central column of boxes lists possible strategies for reducing resource limitation. The right column of boxes and the curved arrows indicate physiological consequences of adjustments listed in the central boxes. (See color plate.)

whereas the pistillate flowers of females, which are fewer in number but with much larger calyces, maintain their carbon fixation capacity for considerably longer and contribute more to their own costs. In dioecious *Aciphylla glaucescens* (Apiaceae), male inflorescences contribute significantly more to the carbon costs of reproduction than do females, owing to a 60% greater area of photosynthetic spines along their larger inflorescences (Hogan *et al.*, 1998). Pistillate flowers have an additional photosynthetic source in the form of green carpels, and, like *S. latifolia* flowers, female inflorescences supply new assimilates throughout fruiting, much longer than for males. These data suggest the potential for floral photosynthesis to offset reproductive costs in both sexes but, in many cases, the bulk of the proximate benefit may be limited to fruit and seed function. It is worth noting that although photosynthesis by reproductive structures is common, a net positive carbon balance is achieved only at early stages of development (Goldstein *et al.*, 1991) or under certain environmental conditions (Cipollini and Levey, 1991).

B. Increased Vegetative Photosynthesis

Studies of ^{14}C translocation show that fruits gain much of their carbon from increased photosynthesis in nearby leaves (e.g., Wardlaw, 1990; Marshall, 1996; Preston, 1999; Fig. 5.2). In fact, there is a considerable evidence that source–sink relations directly alter vegetative photosynthetic capacity (Wardlaw, 1990; Farrar, 1993; Marshall, 1996). For instance, pollination and subsequent fruit maturation increase leaf photosynthesis in females of dioecious *Silene latifolia* (Laporte and Delph, 1996), dioecious *Salix arctica* (Dawson and Bliss, 1993), and *Cannabis sativa* (Dzhaparidze 1969, cited in Dawson and Geber, 1999). These effects, however, are likely to be transitory. For example, McDowell *et al.* (2000) found that photosynthesis on reproductive shoots of *Pinus menzeseii* was 2.5-times greater than that of nonreproductive shoots, but this was limited to the first month after cone initiation. This pattern might suggest that the need for increased carbon gain abates after the construction costs of sinks are largely paid. Several studies have found no evidence for increased local photosynthesis (Evans, 1991; Karlsson, 1994; Obeso *et al.*, 1998). Increased photosynthetic rates can result in water loss via transpiration as more CO_2 is taken up as well as a requirement for additional nitrogen for photosynthetic machinery (Fig. 5.2). The prevalence of enhanced photosynthesis as a strategy to increase reproductive resources is likely to be regulated by the availability of other vital resources, leaving the importance of source-stimulation to overall reproductive costs an open question.

C. Increased Resource Uptake and Resource Use Efficiency

High levels of reproductive investment may not affect rates of assimilation if plants instead respond by increasing their uptake of resources or use them more efficiently (Fig. 5.2). This may be the most likely outcome for gynodioecious species, if fruit production is the primary sink regulating the rates of photosynthesis during reproduction (Poot, 1997; Schultz, 2003 and see preceding). The ancillary cost of increased resource uptake may correspond to construction and maintenance costs of root or leaf production (Fig. 5.2). Increased use efficiency is most likely to occur for nitrogen (N) and water. Photosynthetic nitrogen use efficiency (PNUE) can be achieved by increasing allocation of N to photosynthetic machinery and by maintaining higher intracellular CO_2 concentrations (Fig. 5.2, red arrows). Water use efficiency (WUE) is primarily controlled by stomatal behavior; reduced stomatal aperture increases WUE by reducing transpirational water lost per carbon gained. However, WUE often occurs at the expense of both carbon assimilation (A) and PNUE, because stomatal closure also limits the uptake of CO_2 (Fig. 5.2, blue arrows). Overall, physiological efficiency should reflect a balance between diffusional and biochemical limitations to photosynthesis (Geber and Dawson, 1997).

D. Reabsorption

Some of the resources invested in reproduction are recoverable and this may partially mitigate proximate costs. Any structure that remains attached to the parent plant, rather than being abscised or dispersed at maturity, contains potentially reclaimable nutrients. For example, reabsorption of unremoved nectar occurs in some plant species (e.g., Burquez and Corbet, 1991), and there is evidence that reabsorption of these energy-rich sugars increases fruit quality and seed fitness in an epiphytic orchid (Luyt and Johnson, 2002). A study in *Sidalcea oregana* revealed that in contrast to plant parts shed at maturity (pollen and seeds), ~50–80% of nitrogen and phosphorus is reabsorbed from floral structures that remain attached to the plant (calyces, petals, unfertilized ovules; Ashman, 1994b). There is also potential for the reabsorption of resources from early aborted fruits, although there is little evidence to address it (Obeso, 2002). In contrast to unfertilized ovules, plants are not able to recoup investments from unsuccessful pollen (i.e., left in the flower, lost, or consumed by pollinators). In addition, because pistillate and perfect flowers are maintained longer than staminate ones, sex morphs are likely to differ in their propensity for reabsorption, even from attractive structures. However, benefit of reabsorption should at least offset respiratory costs to translocate the recovered resources (Fig. 5.2).

V. Predictions for Sex-specific Physiology Based on Differential Reproductive Costs

It is difficult to make straightforward predictions of how the sexes should differ physiologically because there are many other life history and ecological traits commonly associated with alternate sex morphs (Sakai and Weller, 1999; Delph, 1999; Dawson and Geber, 1999). For some species, sex-specific life history strategies or spatial segregation may buffer changes to the physiological environment, maintaining homeostasis within photosynthetic organs (Ackerly *et al.*, 2000), such that no physiological sex dimorphism should be expected (see Table 2 in Delph, 1999, for life history traits that may offset reproductive costs). Measurement of integrated rather than instantaneous physiological traits (e.g., using stable isotopes, Dawson *et al.*, 2002), as well as estimates of whole-plant carbon or nutrient budgets (e.g., Laporte and Delph, 1996) are likely to be most informative, but the data are scarce.

In the following section, we make specific predictions of how variation in reproductive investment may be associated with sex-based differences in physiology. Our simplifying assumptions are that sex morphs are not spatially or ecologically segregated (i.e., they occur in microsites of similar quality), and do not differ overall in vegetative size. Thus, our predictions isolate leaf-level physiological differences (and their potential consequences) based solely on the type, amount, and timing of resources invested into reproductive structures (Table 5.1 and Fig. 5.2). We differentiate predictions among sex morphs in the context of particular sexual systems. In particular, we do not make global predictions for how hermaphrodites should differ physiologically from females or males. This is because the balance between pollen and seed production (hence the type and amount of overall reproductive costs) can vary tremendously among hermaphroditic plants, depending on: (1) whether they reproduce within monomorphic or dimorphic populations, (2) whether they are monoecious or perfect flowered, and (3) whether they invest simultaneously or sequentially in both sex functions.

A. Predictions for Females and Males

As previously discussed, females and males differ dramatically in the timing and magnitude of their proximal reproductive costs, with females bearing

Table 5.1   Some Examples of Physiological Mechanisms for Accommodating Proximal Costs of Reproductive Investment by Plants, Grouped by the Major Resource Required

Resource in greatest demand by sinks	Physiological change to minimize resource limitation	Ancillary costs of physiological adjustments
Carbon	a. Increase stomatal conductance to maintain high intracellular CO_2 b. Increase assimilation rates, maximum PS capacity	a. Increased transpirational water loss b. Greater nitrogen requirement for PS machinery
Mineral nutrients (e.g., nitrogen)	a. Increase nutrient uptake b. Increase nutrient resorption c. Increase nutrient use efficiency	a. Root construction & maintenance b. Respiratory cost of translocation c. Increased conductance to maintain higher intracellular CO_2, can increase water loss
Water	a. Increase water uptake b. Increase water use efficiency	a. Root construction & maintenance, increased transpiration b. Reduced stomatal conductance lowers carbon assimilation

the greater total cost in most cases. All else being equal, females should have a greater capacity for resource acquisition via uptake, assimilation (by leaves or reproductive structures) and/or reabsorption than males throughout the reproductive phase, but particularly at fruiting when the bulk of their resource costs are incurred. However, this may or may not be apparent at the leaf-level, particularly if the sexes differ dramatically in vegetative size, age at first reproduction, or frequency of flowering (Delph, 1999).

In terms of resource use efficiency, we predict that males should be more nitrogen use efficient (NUE) than females because of their investment in nitrogen-rich pollen, particularly in cases where nitrogen is in low supply. In fact, the absence of male function may aid females in maintaining higher photosynthetic capacity, by allocating nitrogen to photosynthetic machinery rather than pollen (Poot, 1997; see Fig. 5.2). Differences in NUE may be most pronounced during flowering, when the cost of male function is most evident. In contrast to nitrogen, we expect a stage-specific pattern of water use efficiency (WUE) particularly in cases where floral display size differs between the sexes. Petals are particularly costly in terms of water (Galen *et al.*, 1999). This may select for greater WUE or greater water uptake during flowering for the sex bearing larger petals, longer-lived flowers, or larger daily floral displays. In most temperate dioecious species, males have either more or larger flowers, but female flowers are often longer lived, thus any pattern of sex-specific WUE might be expected. In most cases, we might expect males to be more water use efficient, both because their water costs are greater during flowering, and because females may keep stomates open at the expense of WUE in order to maintain higher rates of carbon assimilation, particularly during fruiting (see Fig. 5.2). Because fleshy fruits are also water costly (Herrera, 2002), females of fleshy-fruited species may: (1) increase WUE from flowering to fruiting, (2) do the bulk of their carbon assimilation during flowering, or (3) depend more on stored resources during fruiting to reduce the water costs of fixing new carbon.

B. Predictions for Hermaphrodites in Monomorphic Sexual Systems: Cosexuality, Monoecy, and Diphasy

Hermaphrodites in monomorphic (hermaphroditic) populations are more appropriately called "cosexuals" because all plants invest in, and reproduce to some extent through, both female and male function (Lloyd, 1976). Because fitness via pollen and seed must be equivalent at the population level, selection should maintain a more or less balanced hermaphroditism among individuals, distinct from the biased allocation patterns expected for hermaphrodites in gender dimorphic (e.g., gynodioecious, or subdioecious) populations (Delph, 2003, and see Section V.C below).

Cosexuals are expected to have a greater total reproductive expenditure than unisexuals, because they produce both types of gametes and are likely to invest more into attractive structures than females or males. Two recent comparative studies predict that the evolution of separate sexes is associated with a reduction in floral display size and/or individual flower size (Miller and Venable, 2003; Vamosi *et al.*, 2003), suggesting higher maintenance and construction costs for cosexuals relative to their unisexual relatives. All else being equal, we predict that cosexuals should have a greater capacity for total resource acquisition (via uptake, assimilation, or reabsorption), or be more resource use efficient than unisexuals throughout flowering and fruiting.

In monoecious species, the production of pistillate and staminate flowers is spatially and often temporally segregated, providing the potential for unique physiological conditions during the bulk of investment into female versus male function. However, floral sex ratios of monoecious species are highly plastic, and determined largely by hormonal signals (e.g., ethylene), which in turn may influence physiological processes.

In the case of complete dichogamy (diphasy), hermaphrodites incur sex-specific reproductive costs sequentially, such that there is no overlap between female and male function. We might predict physiological differences to coincide with reproductive phase – greater resource uptake or assimilation during female relative to male phase. Evidence for physiological differences between sex phases would be particularly useful for understanding the importance of genotype to the evolution of physiological traits, as individuals can be measured at both stages (see preceding). However, phase changes from nonreproductive, to male, then female is almost invariably associated with increases in vegetative size (Vitt, 2001), which may reduce selection for changes in individual leaf functioning, or may favor greater resource acquisition during the male phase to support both reproduction and increased growth or storage.

C. Predictions for Hermaphrodites in Dimorphic Sexual Systems: Gynodioecy and Subdioecy

We can extend the above predictions to situations where the sex functions are incompletely separated among individuals. However, there are two unique issues to bear in mind when considering "intermediate" sexual systems. First, they are often interpreted as transitory states along a pathway between cosexuality and dioecy. Thus, depending on the cause of physiological change (see Section VI), sex morphs may be more similar to each other with respect to physiology, if gender dimorphism has only recently evolved and there has been little time or opportunity for selection. Second, it is critical to note that not all hermaphrodites reproduce equally, and that they are *expected* to bias their reproductive effort in favor of one or the other sex function in the presence of unisexuals. For example, the presence of females in gynodioecious (and subdioecious) populations should increase the male fitness of hermaphrodites, such that they should preferentially invest in traits that enhance pollen dispersal (e.g., pollen production; Lloyd, 1976).

We expect physiological functioning to scale with relative investment (pollen, fruits/seeds, and flowers); resource acquisition and use efficiency should increase with total expenditure, because of the nature of source–sink dynamics (e.g., Reekie and Bazzaz, 1987; Saulnier and Reekie, 1995). Genetic differences among hermaphrodites in the *degree* to which they conditionally adjust their reproductive allocation may be reflected in physiological traits as well. Male-biased hermaphrodites with high gender plasticity are expected to be selectively maintained in, and contribute to the stability of, gynodioecy (Delph, 2003). Genotypes with little gender plasticity may be more prone to physiological stress than plastic genotypes as resource levels change. Hermaphrodites may therefore differ in their physiological responses to environmental heterogeneity according to the degree of gender plasticity – greater emphasis on acquisition and uptake for nonplastic genotypes, and on altered use efficiency with increasing plasticity.

VI. Potential Causes of Sex-specific Physiology

The evolutionary dynamics of plant ecophysiological traits are complex at best (see Ackerly *et al.*, 2000, for recent treatise), complicating questions of when and why morph-specific variation in physiology should occur. Physiological characters, particularly instantaneous traits, do vary substantially, and their tendency to change predictably with environmental conditions (e.g., Dudley, 1996) often begs adaptive explanation. Several hypotheses, both adaptive and nonadaptive, may underlie sexual dimorphism (SD) in physiology. In the next sections, we describe three such hypotheses and the evidence in favor of each.

A. Physiological Differences Reflect Plastic Responses to Contrasting Reproductive Allocation between Sexes

If somatic costs of reproduction trade-off directly with photosynthetic machinery, then differential investment by the sexes should result in SD in physiology. There is little doubt that physiological traits are phenotypically plastic, and do interact with many other morphological and developmental traits. But it is important to discern whether differential reproductive investment per se is the cause of the SD. Under this hypothesis, we expect both inter- and intrasexual differences in physiological traits, depending on which reproductive functions have the greatest effect on physiological functioning, and the extent to which these traits vary within versus between sex phenotypes. Similarly, we expect physiological differences to be stage-specific, reflecting inherent stage specificity of the proximal costs of producing pollen versus seed and fruit.

Evidence testing this hypothesis would ideally combine natural variation in reproductive allocation (e.g., flower, fruit, seed, and pollen production) with patterns observed under manipulated resource investment (e.g., pollinated or not, reproducing or not), and be verified both under controlled and field growth conditions.

B. Selection Modifies Physiological Traits after the Separation of the Sexes to Meet Differential Reproductive Costs

Physiological traits may be targeted by selection, and may diverge in alternate sex morphs. We can use the analogy of separate sexes as unique "environmental contexts" for physiological traits, where disparate reproductive costs create differential resource abundance and distribution that influence the strength and direction of selection.

This hypothesis has several prerequisites – foremost, evidence of genetic variation and heritability of physiological traits (Geber and Dawson, 1997; Arntz and Delph, 2001; Arntz *et al.*, 2002). Given evolutionary potential, this hypothesis assumes that sexual dimorphism in physiology follows, rather than precedes, the separation of the sexes. Phylogenetic studies or sister comparisons of groups differing in sexual system may help elucidate the order in which sexually dimorphic traits evolved (e.g., Miller and Venable, 2003). Artificial selection experiments may also reveal the potential for divergence once separate sexes are established (e.g., Delph *et al.*, 2004). If sex morphs can be artificially selected to vary in a physiological trait of interest, e.g., photosynthetic rate or water use efficiency, one could relate this variation to an estimate of fitness for each sex. The creation of broad intramorph variation in physiology via artificial selection that is *not* entirely explained by variation in reproductive investment would evidence the potential for independent selection on physiology after evolutionary transitions in sexual systems. Morph-specific optima for physiological traits would indicate that selection should favor sexually dimorphic physiology.

C. Physiology Changes as a Correlated Response to Selection on Other Traits (e.g., via Pleiotropy or Linkage)

This hypothesis is somewhat similar to the plasticity hypothesis above, but requires that physiological traits share genes with other fitness-related traits. Unlike the plasticity hypothesis, where sex-specific physiology is expected to be stage (= investment) specific, physiological differences via correlated responses should persist regardless of reproductive stage.

The veracity of this hypothesis may be reflected in estimates of genetic correlations between physiological and morphological traits via standard quantitative genetic protocols (Falconer and Mackay, 1996), or the detection of sex-specific physiology during prereproductive stages. One could also select on various morphological and reproductive traits that are predicted to be associated with physiological diversification, then assess the degree of correlated response. It is important to note that selection on a trait in one sex may produce a response for that same trait in another sex if a significant between-sex genetic correlation exists (Lande, 1980; Eckhart, 1993; Ashman, 1999, 2003; Caruso *et al.*, 2003). Between-sex genetic correlations occur when the expression of genes contributing to quantitative traits, including many fitness-related, reproductive, and physiological traits, are not sex-limited. In the presence of a significant between-sex genetic correlation, both sexes should exhibit a response to selection, even if the trait is targeted in only one sex morph. In the context of sex-specific physiology, a positive between- sex genetic correlation for physiological traits will constrain divergence of sex morphs, despite differences in reproductive cost or correlations among morphological, reproductive, and physiological traits.

VII. Available Data on Sex-specific Physiology

Studies of sex differences in physiology are increasing. Although heavily biased towards dioecious species (Table 5.2), researchers are beginning to explore physiological aspects of reproduction by plants in intermediate sexual systems. We still lack sufficient data to test our predictions, either for the general patterns or specific causes of sexual dimorphism in physiological traits. But the evidence presented in Table 5.2 suggests that simply gathering more data may not reveal a consistent pattern. There are as many studies showing no sexual dimorphism (9 studies) as there are supporting the generally expected pattern of sex-specific physiology (10 studies), with a comparable number of studies finding that sex differences are context- and condition-dependent (9 studies).

One possible reason for inconsistency in pattern is obviously inconsistency of method and experimental design. Each of these studies was carried out under a different set of conditions – some to test the effects of sink strength, others to test the effect of environmental variation on physiological functioning, some using naturally occurring variation, others manipulating condition or context. But probably more important is variation among species in their life histories, and the potential for other traits to buffer or compensate for changes within their leaves.

Several recent studies should be highlighted because they have the potential to provide significant insight into the mechanism by which sex-specific physiology may arise. Caruso *et al.* (2003) found evidence for physiological sex differences in prereproductive female and hermaphrodite *Lobelia siphilitica*. Females exhibited higher rates of assimilation (per area and per mass), higher stomatal conductance, more efficient light harvesting (Fv/Fm), and greater chlorophyll concentrations when compared

Table 5.2 Summary of Available Data on Sex-related Differences in Physiological Traits (SDP) in Gender Dimorphic Taxa. For Each Study, we List: Study Species, Sexual System (D = Dioecious, GD = Gynodioecious, SD = Subdioecious, DPH = Diphasic Hermaphrodite), Location of Data Collection or Source Plants, Sex-specific Patterns of Carbon Assimilation Rate (A) and Stomatal Conductance (Gs), Water Status or Water Use Efficiency (WUE), Nitrogen Use Efficiency (NUE) or Leaf N Content

Species	Sex. Syst.	Location of study	A and Gs	WUE, water status	NUE, leaf N content	References
Expected patterns	D, DPH GD, SD		F > M F ≥ H > M	M > F M, H ≥ F	F > M M, H > F	
Studies showing some expected patterns						
Humulus lupulus	D		F = M	M > F		Dzhaparidze 1969, cited in Dawson and Geber, 1999
Acer negundo	D	Both	F > M	M > F	F > M	Dawson and Ehleringer, 1993; Farquhar et al., 1989; Ward et al., 2002
Baccharis dracunculifolia	D	Field		M > F		Espírito-Santo et al., 2003
Diospyros lotus	D		F > M	M > F		Dzhaparidze 1969, cited in Dawson and Geber, 1999
Pistacia mutica	D			M > F		Dzhaparidze 1969, cited in Dawson and Geber, 1999
Populus sosnowskyi	D		F > M			Chrelashvili and Dzhaparidze 1950 cited in Dawson and Geber, 1999
Salix arctica	D	Field	F > M	M > F	M > F	Dawson and Bliss, 1989, 1993; Jones et al., 1999
Salix reticulata	D	Field	F > M	M > F		Dawson, 1990
Lobelia siphilitica	GD	Controlled conditions	F > H	F = H		Caruso et al., 2003
Spinacia spp.	SD	Controlled conditions	F = H = M	M > F and H		Vitale and Freeman, 1985

(*Continued*)

Table 5.2 Summary of Available Data on Sex-related Differences in Physiological Traits (SDP) in Gender Dimorphic Taxa. For Each Study, we List: Study Species, Sexual System (D = Dioecious, GD = Gynodioecious, SD = Subdioecious, DPH = Diphasic Hermaphrodite), Location of Data Collection or Source Plants, Sex-specific Patterns of Carbon Assimilation Rate (A) and Stomatal Conductance (Gs), Water Status or Water Use Efficiency (WUE), Nitrogen Use Efficiency (NUE) or Leaf N Content—cont'd

Species	Sex. Syst.	Location of study	A and Gs	WUE, water status	NUE, leaf N content	References
Studies opposing expected patterns						
Phoradendron juniperinum	D	Field	A: M > F g: F = M	M > F	F = M	Marshall *et al.*, 1993
Populus tremuloides	D		A: M > F g: F > M	M > F	M > F	Sakai *et al.*, unpublished, cited in Dawson and Geber, 1999
Silene latifolia	D	Controlled conditions	M > F	F = M	F = M	Gehring and Monson, 1994; Laporte and Delph, 1996
Arisaema triphyllum	DPH	Controlled conditions	M > F			Vitt, 2001
Studies showing context- or condition-dependent patterns						
Atriplex spp.	D	Field		M > F at frt		Freeman and McArthur, 1982
Cannabis sativa	D	Controlled conditions	M > F at flw; F > M at frt	M > F		Dzhaparidze, 1969, cited in Dawson and Geber, 1999
Ilex aquifolium	D	Controlled conditions		F > M in dry shaded conditions	F = M	Retuerto *et al.*, 2000; Obeso and Retuerto, 2002
Pistacia lentiscus	D	Both	M > F in dry conditions F = M in watered conditions	Water uptake: M > F WUE: F > M	M > F	Jonasson *et al.*, 1997; Correia and Diaz Barradas, 2000

Species		Conditions				Reference
Rubus chamaemorus	D	Field			M > F	Ågren, 1988
Siparuna grandiflora	D	Field	M = F at flw; M > F at frt		M > F	Nicotra, 1997
Fragaria virginiana	GD	Field gardens	H > F at flw; F = H at frt	F > H at flw; H > F in HR at Frt		Case *et al.*, unpublished
Phacelia linearis	GD	Controlled conditions			F > H at low nutrients; H>F at high nutrients	Eckhart and Chapin, 1997
Sidalcea hirtipes	GD	Field	H > F esp. at low Ci	H > F at low Ci	F = H	Schultz, 2003
Studies showing no SDP						
Aciphylla glaucescens	D	Field	F = M		F = M	Hogan *et al.*, 1998
Ephedra viridis	D			F = M		Freeman *et al.*, 1976
Hesperochloa kingii	D			F = M		Fox and Harrison, 1981
Nyssa sylvatica	D	Field	F = M		M = F	Cipollini and Stiles, 1991
Rumex acetosella	D			F = M		Zimmerman and Lechowicz, 1982; Houssard, *et al.*, 1992
Thalictrum ssp.	D		F = M		F = M	Melampy, 1981
Plantago lanceolata	GD	Controlled conditions	F = H		F = H	Poot *et al.*, 1996
Wurmbea dioica ssp. alba	GD	Field		F = H	F = H	Case and Barrett, 2001
Sidalcea oregana ssp. spicata	GD	Field			F = H	Ashman, 1991

with hermaphrodites. They also detected a significant positive between-sex genetic correlations for chlorophyll concentration, specific leaf area, leaf size, and leaf mass. Studies of gynodioecious *Plantago lanceolata* showed genotypic effects, but no consistent physiological differences among clones of hermaphrodite, male sterile, and intermediate sex types at any reproductive stage and under varying resource conditions (Poot *et al.,* 1996, 1997; Poot, 1997). These patterns suggest that sex-specific physiology in this case is not solely a plastic response to differences in sex allocation, and may not evolve entirely independently between the sexes.

Both genotype and reproductive investment appears to influence the expression of physiological traits during the sex phases of diphasic *Arisaema triphyllum* (Vitt, 2001). Male phase plants have greater rates of photosynthesis per area than the same plants during the female phase, but females have greater total carbon assimilation, because reproduction as a female is associated with greater vegetative size. Plants in male phase may maintain higher rates of photosynthesis because they rely more on current photosynthates to support reproduction and growth, while females, with greater leaf area and larger storage organs, are able to acquire more carbon overall to support reproduction as well as storage for future reproductive episodes. These patterns were consistent within genotypes across sex phases, despite substantial variation among genotypes. Further investigations in gender-switching species would be helpful for providing insight into the influence of genotype on the relation between physiology and reproductive costs, as it is often not possible to hold genotype, size, morphology, and growth conditions completely constant across the sexes of gynodioecious or dioecious species.

Laporte and Delph (1996) found greater photosynthesis rates by males than females in *Silene latifolia,* and that photosynthesis rates among females were consistent with whole-plant source–sink ratios such that photosynthesis is upregulated by fruit production. Thus, females can respond plastically to variation in reproductive investment, yet males achieve a greater total carbon balance than females. To further this investigation, Delph (unpublished) has undertaken a multigeneration selection experiment to assess the response of physiological traits to artificial selection on reproductive traits. Selection on floral characters resulted in direct responses in physiology, suggesting the potential for physiology to evolve via correlated responses to selection on reproductive investment.

Nicotra (1997) explored the physiological consequences of reproduction in field populations of dioecious *Siparuna grandiflora.* She contrasted physiological traits of two sets of reproductive and nonreproductive plants at both flowering and during fruiting stages: intact versus deflowered adults, and cuttings of same age plants that either had or had not begun to flower under field conditions. Flower removal substantially increased photosynthetic capacity for females at both reproductive stages, but enhanced photosynthesis only during the late season (i.e., during female fruiting) in males. This result suggests that reproduction reduces photosynthesis in reproducing plants of both sexes, but because the effect was seen earlier in females, they have a capacity to respond more quickly to resource manipulation than do males. In her second data set, flowering versus nonflowering females had similar photosynthetic capacities, while reproductive investment in males increased photosynthesis late in the reproductive season when compared to nonflowering males.

Overall, the only significant difference between sexes of *S. grandiflora* that were reproductively active was that flowering males had greater leaf N content than flowering females (Nicotra, 1997). However, leaf N was differentially affected by reproductive status between the sexes. Flower removal was associated with decreased leaf N for males, but increased leaf N for females, while prereproductive plants of both sexes had greater leaf N than those producing flowers and fruits. Change in leaf N corresponded to increased PNUE and photosynthesis only for females, suggesting that the excess N in male leaves took the form of inactive N, perhaps reflecting nitrogen slated for allocation reproduction. Taken together, she concluded that males probably rely more on current photosynthates for reproductive resources and exact a greater N cost of reproduction, while females tap into stored resources, are able to acclimate their physiological processes more to changes in resource allocation, and overall achieve a similar carbon balance, despite a greater cost of reproduction.

Case and colleagues (unpublished) measured rates of photosynthesis, stomatal conductance and water use efficiency in females and hermaphrodites of *Fragaria virginiana* in an experimental context. They manipulated both soil resource availability and sink strength (hand-pollinated versus flower removal between clones), and assessed physiological responses of both sexes during flowering and fruiting. At flowering, hermaphrodites exhibited significantly higher rates of photosynthesis and stomatal conductance, but lower WUE than females, regardless of soil resource availability. No detectable effect of fruit production was found on any of these traits at fruiting; females and hermaphrodites did not differ in assimilation or conductance, and hermaphrodites were more water use efficient, but only under high resource conditions.

Taken together, the available data suggests that physiological sex differentiation is unlikely to be caused primarily by plastic responses to variation in reproductive investment, leaving direct selection and pleiotropy as more likely causes. Future research should focus on experiments that can distinguish among these potential mechanisms, and perhaps identify other factors (e.g., genetics, life history, ecology) that make one of these mechanisms more likely to cause (or hinder) physiological SD in a particular plant group. For example, the nature of sex determination (e.g., whether sex is determined by nuclear versus cytoplasmic elements, whether male sterility is dominant versus recessive) may affect the likelihood and strength of between-sex genetic correlations, consequently the potential for SD. Both theoretical and empirical advances in this area would provide great insight, and perhaps uncover some hidden patterns in the available data on sex-specific physiology (Table 5.2).

VIII. Recommendations for Future Study

Female plants incur greater reproductive costs than male plants. Beyond this, few general patterns emerge from data on sex-specific physiology, morphology, or life history. If our goal is to understand how and whether sexes should differ physiologically, and to uncover the proximate causes of physiological differentiation, we should start by engaging in more "wholistic" studies. We should attempt to integrate all available evidence for sexual dimorphism – whether reproductive, vegetative, physiological, ecological, or phenological – rather than segregate data by type or scale. At the same time, measures should be taken so that the effects of context and/or condition can be meaningfully evaluated. We should also pay more attention to intermediate sexual systems, where the scale and history of differentiation may be reduced, providing clues as to the origins and selective pressures favoring sex-specific physiology.

Acknowledgments

We gratefully acknowledge insightful comments, helpful discussion, and unpublished data from L. Delph and C. Caruso. One anonymous reviewer provided additional helpful comments on a draft of this manuscript. This research was supported in part by NSF DEB-9903802 to TLA.

References

Ackerly D. D., Dudley S. A., Sultan S. E., Schmitt J., Coleman J. S., Linder C. R., Sandquist D. R., Geber M. A., Evans A. S., Dawson T. E. and Lechowicz M. J. (2000) The evolution of plant ecophysiological traits: recent advances and future directions. *BioScience* **50**: 979–995.

Ågren J. (1988) Sexual differences in biomass and nutrient allocation in the dioecious *Rubus chamaemorus. Ecology* **69**: 962–973.

Ågren J., Danell K., Elmqvist T., Ericson L. and Hjalten J. (1999) Sexual dimorphism and biotic interactions. In *Gender and Sexual Dimorphism in Flowering Plants* (Geber M. A., Dawson T. E. and Delph L. F. eds.) pp. 217–246, Springer-Verlag, New York.

Alonso C. and Herrera C. M. (2001) Neither vegetative nor reproductive advantages account for high frequency of male-steriles in southern Spanish gynodioecious *Daphne laureola* (Thymelaeaceae). *American Journal of Botany* **88**: 1016–1024.

Aronne G., Wilcock C. C. and Pizzolongo. P. (1993) Pollination biology and sexual differentiation of *Osyris alba* (Santalaceae) in the Mediterranean region. *Plant Systematics and Evolution* **188**: 1–16.

Arntz A. M. and Delph L. F. (2001) Pattern and process: evidence for the evolution of photosynthetic traits in natural populations. *Oecologia* **127**: 455–467.

Arntz A. M., Vozar E. M. and Delph L. F. (2002) Serial adjustments in allocation to reproduction: effects of photosynthetic genotype. *International Journal of Plant Sciences* **163**: 591–597.

Ashman T.-L. (1991) The factors governing the maintenance of gynodioecy in *Sidalcea oregana* ssp. *spicata* (Malvaceae). Ph.D. dissertation, University of California, Davis, CA.

Ashman T.-L. (1994a) Reproductive allocation in hermaphrodite and female plants of *Sidalcea oregana* ssp. *spicata* (Malvaceae) using four currencies. *American Journal of Botany* **81**: 433–438.

Ashman T.-L. (1994b) A dynamic perspective on the physiological cost of reproduction in plants. *American Naturalist* **144**: 300–316.

Ashman T.-L. (1999) Determinants of sex allocation in a gynodioecious wild strawberry: implications for the evolution of dioecy and sexual dimorphism. *Journal of Evolutionary Biology* **12**: 648–661.

Ashman T.-L. (2002) The role of herbivores in the evolution of separate sexes from hermaphroditism. *Ecology* **83**: 1175–1184.

Ashman T.-L. (2003) Constraints on the evolution of males and sexual dimorphism: Field estimates of genetic architecture of reproductive traits in three populations of gynodioecious *Fragaria virginiana*. *Evolution* **57(9)**: 2012–2025

Ashman T.-L., Cole D. and Bradburn M. (2004) Sex-differential resistance and tolerance to herbivory in a gynodioecious wild strawberry: Implications for floral and sexual system evolution. *Ecology* **85**: 2550–2559.

Ashman T.-L. and Stanton M. L. (1991) Seasonal variation in pollination dynamics of sexually dimorphic *Sidalcea oregana* ssp. *spicata* (Malvaceae). *Ecology* **72**: 993–1003.

Ashman T.-L. and Schoen D. J. (1994) How long should flowers live? *Nature* **371**: 788–791.

Ashman T.-L. and Schoen D. J. (1997) The cost of floral longevity in *Clarkia tembloriensis*: An experimental investigation. *Evolutionary Ecology* **11**: 289–300.

Assouad M. W., Dommee B. C., Lumaret R. and valdeyron G. (1978) Reproductive capacities in the sexual forms of the gynodioecious species *Thymus vulgaris* L. *Biological Journal of the Linnean Society* **77**: 29–39.

Bazzaz F. A., Carlson R. W. and Harper J. L. (1979) Contribution to the reproductive effort by photosynthesis of flowers and fruits. *Nature* **279**: 554–555.

Belhassen E., Trabaud L., Couvet D. and Gouyon P. H. (1989) An example of nonequilibrium processes: gynodioecy of *Thymus vulgaris* L. in burned habitats. *Evolution* **43**: 662–667.

Boecklen W. J. and Hoffman M. T. (1993) Sex-biased herbivory in *Ephedra trifurca*: the importance of sex-by-environment interactions. *Oecologia* **96**: 49–55.

Burquez A. and Corbet S. A. (1991) Do flowers reabsorb nectar? *Functional Ecology* **5**: 369–379.

Carroll S. B. and Delph L. F. (1996) The effects of gender and plant architechture on allocation to flowers in dioecious *Silene latifolia* (Caryophyllaceae). *International Journal of Plant Sciences* **157**: 493–500.

Caruso C. M., Maherali H. and Jackson R. B. (2003) Gender-specific floral and physiological traits: implications for the maintenance of females in gynodioecious *Lobelia siphilitica*. *Oecologia* **135**: 524–531.

Case A. L. and Barrett S. C. H. (2001) Ecological differentiation of combined and separate sexes of Wurmbea dioica (Colchicaceae) in sympatry. *Ecology* **82**: 2601–2616.

Charlesworth D. (1999) Theories of the evolution of dioecy. In *Gender and Sexual Dimorphism in Flowering Plants* (Geber M. A., Dawson T. E and Delph L.F. eds.) pp. 33–60, Springer-Verlag, New York.

Charnov E. L. (1982) *The Theory of Sex Allocation*. Princeton University Press, Princeton.

Cipollini M. L. and Levey D. J. (1991) Why some fruits are green when they are ripe – carbon balance in fleshy fruits. *Oecologia* **88**: 371–377.

Cipollini M. L. and Stiles E. W. (1991) Costs of reproduction in *Nyssa sylvatica*: sexual dimorphism in reproductive frequency and nutrient flux. *Oecologia* **86**: 585–593.

Connor H. E. (1973) Bredding systems in Cortaderia (Gramineae). *Evolution* **27**: 663–678.

Correia O. and Diaz Barradas M. C. (2000) Ecophysiological differences between male and female plants of *Pistacia lentiscus* L. *Plant Ecology* **149**: 131–142.

Crawford R. M. M. and Balfour J. (1983) Female predominat sex ratios and physiological differentiation in arctic willows. *Journal of Ecology* **71**: 419–160.

Dawson T. E. (1990) Spatial and physiological overlap of three co-occurring alpine willows. *Functional Ecology* **4**: 13–25.

Dawson T. E. and Bliss L. C. (1989) Patterns of water use and the tissue water relations in the dioecious shrub, *Salix arctica*: the physiological basis for habitat partitioning between the sexes. *Oecologia* **79**: 332–343.

Dawson T. E. and Bliss L. C. (1993) Plants as mosaics: leaf-, ramet-, and gender-level variation in the physiology of the dwarf willow, *Salix arctica*. *Functional Ecology* **7**: 293–304.

Dawson T. E. and Ehleringer J. R. (1993) Gender-specific physiology, carbon isotope discrimination, and habitat distribution in boxelder, *Acer negundo*. *Ecology* **74**: 798–815.

Dawson L. E. and Geber M. A. (1999) Dimorphism in physiology and morphology. In *Gender and Sexual Dimorphism in Flowering Plants* (Geber M. A., Dawson T. E. and Delph L. F. eds.) pp. 175–215, Springer-Verlag, New York.

Dawson T. E., Mambelli S., Plamboeck A. H., Templer P. H. and Tu K. P. (2002) Stable isotopes in plant ecology. *Annual Review of Ecology and Systematics* **33**: 507–559.

Delph L. F. (1990) Sex differential resource allocation patterns in the subdioecious shrub *Hebe subalpina*. *Ecology* **71**: 1342–1351.

Delph L. F. (1999) Sexual dimorphism in life history. In *Gender and Sexual Dimorphism in Flowering Plants.* (Geber M. A., Dawson T. E. and Delph L. F. eds.) pp. 149–173, Springer-Verlag, New York.

Delph L. F. (2003) Sexual dimorphism in gender plasticity and its consequences for breeding system evolution. *Evolution and Development* **5**: 34–39.

Delph L. F., Gehring J. L., Frey F. M., Arntz A. M. and Levri M. (2004) Genetic constraints on floral evolution in a sexually dimorphic plant revealed by artificial selection. *Evolution* **58**: 1936–1946.

Dudley S. A. (1996) The response to differing selection on plant physiological traits: Evidence for local adaptation. *Evolution* **50**: 103–110.

Eckhart V. M. (1992) Resource compensation and the evolution of gynodioecy in *Phacelia linearis* (Hydrophyllaceae). *Evolution* **46**: 1313–1328.

Eckhart V. M. (1993) Do hermaphrodites of gynodioecious *Phacelia linearis* (Hydrophyllaceae) trade off seed production to attract pollinators? *Biol J Linn Soc* **50**: 47–63

Eckhart V. M. (1999) Sexual dimorphism in flowers and inflorescences. In *Gender and Sexual Dimorphism in Flowering Plants* (Geber M. A., Dawson T. E. and Delph L. F. eds.) pp. 123–148, Springer-Verlag, New York.

Eckhart V. M. and Chapin F. S. (1997) Nutrient sensitivity of the cost of male function in gynodioecious *Phacelia linearis*. *American Journal of Botany* **84**: 1092–1098.

Eckhart V. M. and Seger J. (1999) Phenological and developmental costs of male sex function in hermaphroditic plants. In *Life History Evolution in Plants* (Vuorisalo T. O. and Mutikainen P. K. eds.) pp. 195–213, Kluwer, Dordrecht.

Ehrlen J. and van Groenendael J. (2001) Storage and the delayed costs of reproduction in the understorey perennial *Lathyrus vernus*. *Journal of Ecology* **89**: 237–246.

Espírito-Santo M. M., Madeira B. G., Neves F. S., Faria M. L., Fagundes M. and Wilson Fernandes G. (2003) Sexual differences in reproductive phenology and their consequences for the demography of *Baccharis dracunculifolia* (Asteraceae), a dioecious tropical shrub. *Annals of Botany* **91**:13–19.

Evans A. S. (1991) Whole-plant responses of *Brassica ampestris* (Cruciferae) to altered sink-source relations. *American Journal of Botany* **78**: 394–400.

Falconer D. S. and Mackay T. F. C. (1996) *Introduction to Quantitative Genetics.* Longman Group Ltd., Essex, England.

Farquhar G. D., Ehleringer J. R. and Hubick K. T. (1989) Carbon isotope discrimination and photosynthesis. *Annual Review of Plant Physiology and Plant Molecular Biology* **40**: 503–547.

Farrar J. F. (1993) Sink strength: what is it and how do we measure it? *Plant Cell and Environment* **16**: 1013–1046.

Fleming T. H., Maurice S., Buchmann S. L. and Tuttle M. D. (1994) Reproductive biology and relative male and female fitness in a trioecious cactus, *Pachycereus pringlei* (Cactaceae). *American Journal of Botany* **81**: 858–867.

Fox J. F. and Harrison A. T. (1981) Habitat assortment of sexes and water balance in a dioecious grass. *Oecologia* **49**: 233–235.

Freeman D. C. and McArthur E. D. (1982) A comparison of twig water stress between males and females of six species of desert shrubs. *Forest Science* **28**: 304–308.

Freeman D. C., Klikoff L. G. and Harper K. T. (1976) Differential resource utilization by the sexes of dioecious plants. *Science* **193**: 597–599.

Galen C. (1993) Cost of reproduction in Polemonium viscosum – phenotypic and genetic approaches. *Evolution* **47**: 1073–1079.

Galen C., Dawson T. E. and Stanton M. L. (1993) Carpels as leaves – meeting the carbon cost of reproduction in an alpine buttercup. *Oecologia* **95**: 187–193.

Galen C., Sherry R. A. and Carroll A. B. (1999) Are flowers physiological sinks or faucets? Costs and correlates of water use by flowers of *Polemonium viscosum. Oecologia* **118**: 461–470.

Geber M. A. and Dawson T. E. (1997) Genetic variation in stomatal and biochemical limitations to photosynthesis in the annual plant, *Polygonum arenastrum. Oecologia* **109**: 535–546.

Geber M. A., Dawson T. E. and Delph L. F. (editors) (1999) *Gender and Sexual Dimorphism in Flowering Plants.* Springer-Verlag, New York.

Gehring J. L. and Monson R. K. (1994) Sexual differences in gas exchange and response to environmental stress in dioecious *Silene latifolia* (Caryophyllaceae). *American Journal of Botany* **81**: 166–174.

Goldstein G, Sharifi M. R., Kohorn L. U., Lighton J. R. B., Shultz L. and Rundel P. W. (1991) Photosynthesis by inflated pods of a desert shrub, *Isomeris arborea. Oecologia* **85**: 396–402

Hemborg A. M. and Karlsson P. S. (1999) Sexual differences in biomass and nutrient allocation of first-year *Silene dioica* plants. *Oecologia* **118 (4)**: 453–460.

Hermanutz L. A. and Innes D. J. (1994) Gender variation in *Silene acualis* (Caryophyllaceae). *Plant Systematics and Evolution* **191**: 69–81.

Herrera C. M. (2002) Seed dispersal by vertebrates. In *Plant-Animal Interaction: An Evolutionary Approach* (Herrera C. M. and Pellmyr O. eds.) pp. 185–208, Blackwell Science Ltd., Malden, MA.

Hogan K. P., García M. B. Cheeseman J. M. and Loveless M. D. (1998) Inflorescence photosynthesis and investment in reproduction in the dioecious species *Aciphylla glaucescens* (Apiaceae). *New Zealand Journal of Botany* **36**: 653–660.

Horovitz A. (1980) Gynodioecy as a possible population strategy for increasing reproductive output. *Theoretical and Applied Genetics* **57**: 11–15.

Houssard C., Escarre J. and Vartanian N. (1992) Water stress effects on successional populations of the dioecious herb, *Rumex acetosella* L. *New Phytologist* **120**: 551–559.

Jackson L. L. and Dewald C. L. (1994) Predicting the evolutionary consequences of greater reproductive effort in *Tripsacum dactyloides,* a perennial grass. *Ecology* **75**: 627–641.

Jonasson S., Medrano H. and Flexas J. (1997) Variation in leaf longevity of *Pistacia lentiscus* and its relationship to sex and drought stress inferred from leaf Δ^{13}C. *Functional Ecology* **11**: 282–289.

Jones A. and Burd M. (2001) Vegetative and reproductive variation among unisexual and hermaphroditic individuals of *Wurmbea dioica* (Colchicaceae). *Australian Journal of Botany* **49**: 603–609.

Jones M. H., Macdonald S. E. and Henry G. H. R. (1999) Sex- and habitat-specific responses of a high arctic willow, *Salix arctica,* to experimental climate change. *Oikos* **87**: 129–138.

Jordano P. (1992) Fruits and frugivory. In *Seeds: The Ecology of Regeneration in Plant Commuities.* (Fenner M. ed.) pp. 105–156, Commonwealth Agricultural Bureau International, Wallingford.

Jurik T. W. (1983) Reproductive effort and CO_2 dynamics of wild strawberry populations. *Ecology* **64**: 1329–1342.

Karlsson P. S. (1994) Photosynthetic capacity and photosynthetic nutrient use efficiency of Rhododendron lapponicum leaves as related to nutrient status, leaf age, and branch reproductive status. *Functional Ecology* 8: 694–700.

Kesseli R. and Jain.S. K. (1984) An ecological genetic study of gyndioecy in *Limnanthes douglasii* (Limnanthaceae). *American Journal of Botany* **71**: 775–786.

Kohn J. R. (1989) Sex ratio, seed production, biomass allocation, and the cost of male function in *Cucurbita foetidissima* HBK (Cucurbitaceae). *Evolution* **43**: 1424–1434.

Krohne D. T., Baker I. and Baker H. G. (1980) The maintenance of the gynodioecious breeding system in *Plantago lanceolata* L. *The American Midland Naturalist* **103**: 269–279.

Lande R. (1980) Sexual dimorphism, sexual selection, and adaptation in polygenic characters. *Evolution* **34**: 292–305.

Laporte M. M. and Delph L. F. (1996) Sex-specific physiology and source sink relations in the dioecious plant *Silene latifolia. Oecologia* **106**: 63–72.

Lloyd D. G. (1976) The transmission of genes via pollen and ovules in gyodioecious angiosperms. *Theoretical Population Biology* 9: 299–316.

Lloyd D. G. and Myall A. J. (1976) Sexual dimorphism in *Cirsium arvense* (L.) Scop. *Annals of Botany* **40**: 115–123.

Luyt R. and Johnson S. D. (2002) Postpollination nectar reabsorption and its implications for fruit quality in an epiphytic orchid. *Biotropica* **34**: 442–446.

Marshall C. (1996) Sectorality and physiological organisation in herbaceous plants: an overview. *Vegetatio* **127**: 9–16.

Marshall J. D., Dawson T. E. and Ehleringer J. R. (1993) Gender-related differences in gas exchange are not related to host quality in the xylem-tapping mistletoe *Phoradendron juniperinum* (Viscaceae). *American Journal of Botany* **80**: 641–645.

McDowell S. C. L., NcDowell N. G., Marshall J. D. and Hultine K. (2000) Carbon and nitrogen allocation to male and female reproduction in Rocky Mountain Douglas fir (*Pseudotsuga menziesii* var. *galuca*, Pinaceae). *American Journal of Botany* **87**: 539–546.

Melampy M. N. (1981) Sex-linked niche differentiation in two species of *Thalictrum. American Midland Naturalist* **106**: 325–334.

Miller J. S. and Venable D. L. (2003) Floral morphometrics and the evolution of sexual dimorphism in *Lycium* (Solanaceae). *Evolution* **57**: 74–86.

Molina-Freaner F. and Jain S. K. (1992) Female requencies and fitness components between sex phenotypes among gynodioecious populations of the colonizing species *Trifolium hirtum* All. in California. *Oecologia* **92**: 279–286.

Nicotra A. B. (1997) Functional ecology of dioecy in *Siparuna grandiflora*, a tropical understory shrub. Ph.D. dissertation, University of Connecticut, Storrs, CT.

Obeso J. R. (2002) The costs of reproduction in plants. *New Phytologist* **155**: 321–348.

Obeso J. R., Alvarez-Santullano M. and Retuerto R. (1998) Sex ratios, size distributions, and sexual dimorphism in the dioecious tree Ilex aquifolium (Aquifoliaceae). *Am J Bot* **85**: 1602–1608.

Obeso J. R. and Retuerto R. (2002) Sexual dimorphism in holly (*Ilex aquifolium*): cost of reproduction, sexual selection, or physiological differentiation? *Revista Chilena de Historia Natural* **75**: 67–77.

Pettersson M. W. (1992) Advantages of being a specialist female in gynodioecious *Silene vulgaris* L. (Caryophyllaceae). *American Journal of Botany* **79**: 1389–1395.

Poot P. (1997) Reproductive allocation and resource compensation in male-sterile and hermaphroditic plants of *Plantago lanceolata* (Plantaginaceae). *American Journal of Botany* **84**: 1256–1265.

Poot P., Pilon J. and Pons T. L. (1996) Photosynthetic characteristics of leaves of malesterile and hermaphrodite sex types of *Plantago lanceolata* grown under conditions of contrasting nitrogen and light availabilities. *Physiologia plantarum* 98: 780–790.

Poot P., van den Broek T., van Damme J. M. M. and Lambers H. (1997) A comparison of the vegetative growth of male-sterile and hermaphroditic lines of *Plantago lanceolata* in relation to N supply. *New Phytologist* 135: 429–437.

Preston K. A. (1999) Can plasticity compensate for architechtural constraints on reproduction? Patterns of seed production and carbohydrate translocation in *Perilla frutescens*. *Journal of Ecology* 87: 697–712.

Primack R. B. (1978) Evolutionary aspects of wind pollination in the genus *Plantago* (Plantaginaceae). *New Phytologist* 81: 449–458.

Primack R. B. (1985) Longevity of individual flowers. *Annual Review of Ecology and Systematics* 16: 15–37.

Pyke G. H. (1991) What does it cost a plant to produce floral nectar? *Nature* 350: 58–59.

Ramsey M. (1997) No evidence for demographic costs of seed production in the pollen-limited perennial herb *Blandfordia grandiflora* (Liliaceae). *International Journal of Plant Sciences* 158: 785–793.

Ramsey M. and Vaughton G. (2002) Maintenance of gynodioecy in *Wurmbea biglandulosa* (Cholicaceae): gender differences in seed production and progeny success. *Plant Systematics and Evolution* 232.

Reekie E. G. and Bazzaz F. A. (1987) Reproductive effort in plants. 1. Carbon allocation to reproduction. *American Naturalist* 129: 876–896.

Renner S. S. and Ricklefs R. E. (1995) Dioecy and its correlates in the flowering plants. *American Journal of Botany* 82: 596–606.

Retuerto R., Fernandez Lema B., Rodriguez Roiloa S. and Obeso J. R. (2000) Gender, light, and water effects in carbon isotope discrimination and growth rates in the dioecious tree *Ilex aquifolium*. *Functional Ecology* 14: 529–537.

Roulston T. H., Cane J. H. and Buchmann S. L. (2000) What governs protein content of pollen: Pollinator preferences, pollen–pistil interactions, or phylogeny? *Ecological Monographs* 70: 617–643.

Sakai A. K. and Weller S. G. (1991) Ecological aspects of sex expression in subdioecious Schiedea globosa (Caryophyllaceae). *American Journal of Botany* 78: 1280–1288.

Sakai A. K. and Weller S. G. (1999) Gender and sexual dimorphism in flowering plants: a review of terminology, biogeographic patterns, ecological correlates, and phylogenetic approaches. In *Gender and Sexual Dimorphism in Flowering Plants* (Geber M. A., Dawson T. E. and Delph L. F. eds.) pp. 1–31, Springer-Verlag, New York.

Sakai A. K., Weller S. G., Chen M.-L., Chou S.-Y. and Tasanont T. (1997) Evolution of gynodioecy and the maintenance of females: the role of inbreeding depression, outcrossing rates, and resource allocation in *Schiedea adamantis* (Caryophyllaceae). *Evolution* 51: 724–736.

Saulnier T. P. and Reekie E. G. (1995) Effect of reproduction on nitrogen allocation and carbon gain in *Oenothera biennis*. *Journal of Ecology* 83: 23–29.

Schultz S. T. (2003) Sexual dimorphism in gynodioecious *Sidalcea hirtipes* (Malvaceae). I. Seed, fruit and ecophysiology. *International Journal of Plant Sciences* 164: 165–173.

Shykoff J. A. (1988) Maintenance of gynodioecy in *Silene acaulis* (Caryophyllaccac): stage-specific fecundity and viability selection. *American Journal of Botany* 75: 844–850.

Shykoff J. A., Kolokotronis S.-O., Collin C. L. and López-Villavincencio M. (2003) Effects of male sterility on reproductive traits in gynodioecious plants: a metaanalysis. *Oecologia* 135: 1–9.

Stevens D. P. (1988) On the gynodioecious polymorphism in *Saxifraga granulata* L. (Saxifragaceae). *Biological Journal of the Linnean Society* 35: 15–28.

Vamosi J. C., Otto S. P. and Barrett S. C. H.(2003) Phylogenetic analysis of the ecological correlates of dioecy in angiosperms. *Journal of Evolutionary Biology* 16: 1006–1018.

Vaughton G. and Ramsey M. (2002) Evidence of gynodioecy and sex ratio variation in *Wurmbea biglandulosa* (Colchicaceae). *Plant Systematics and Evolution* **232**: 167–179.

Vitale J. J. and Freeman D. C. (1985) Secondary sex characteristics in *Spinacia oleracea* L.: quantitative evidence for the existence of at least three sexual morphs. *American Journal of Botany* **72**: 1061–1066.

Vitt P. (2001) Gender-related differences in gas exchange rates in the gender-switching species *Arisaema triphyllum. Rhodora* **103**: 387–404.

Ward J. K., Dawson T. E. and Ehleringer J. R. (2002) Responses of *Acer negundo* genders to interannual differences in water availbility determined from carbon isotope ratios of tree ring cellulose. *Tree Physiology* **22**: 339–346.

Wardlaw I. F. (1990) The control of carbon partitioning in plants. *New Phytologist* 116: 341–381.

Webb C. J. (1999) Empirical studies: evolution and maintenance of dimorphic breeding systems. In *Gender and Sexual Dimorphism in Flowering Plants* (Geber M. A., Dawson T. E. and Delph L.F. editors.) pp. 61–95, Springer-Verlag, New York.

Weiblen G. D., Oyama R. K. and Donoghue M. J. (2000) Phylogenetic analysis of dioecy in monocotyledons. *American Naturalist* **155**: 46–58.

Wolfe L. M. and Shmida A. (1997) The ecology of sex expression in a gynodioecious Isreali desert shrub (*Orchradenus baccatus*). *Ecology* **78**: 101–110.

Zangerl A. R. and Berenbaum M. R. (1990) Furanocoumarin induction in wild parsnip – genetics and population variation. *Ecology* **71 (5)**: 1933–1940.

Zimmerman J. K. and Lechowicz M. J. (1982) Responses to moisture stress in male and female plants of *Rumex acetosella. Oecologia* **53**: 305–309.

Appendix

References used in the production of Fig. 5.1: Connor, 1973; Lloyd and Myall, 1976; Assouad *et al.,* 1978; Primack, 1978; Horovitz, 1980; Krohne *et al.,* 1980; Kesseli and Jain, 1984; Shykoff, 1988; Stevens, 1988; Belhassen *et al.,* 1989; Kohn, 1989; Delph, 1990; Sakai and Weller, 1991; Eckhart, 1992; Molina-Freaner and Jain, 1992; Pettersson, 1992; Aronne *et al.,* 1993; Ashman, 1994a; Fleming *et al.,* 1994; Hermanutz and Innes, 1994; Sakai *et al.,* 1997; Wolfe and Shmida, 1997; Ashman, 1999; Alonso and Herrera, 2001; Jones and Burd, 2001; Ramsey and Vaughton, 2002; Vaughton and Ramsey, 2002; Schultz, 2003; Ashman (unpublished data).

6

Time of Flowering, Costs of Reproduction, and Reproductive Output in Annuals

Tadaki Hirose, Toshihiko Kinugasa, Yukinori Shitaka

I. Introduction

Annual plants complete their growth and reproduction within a year (Harper, 1977). Summer annuals germinate in spring, grow vegetatively through summer, then flower and set seeds before the onset of winter. Winter annuals germinate in autumn, remain vegetative often making a rosette during wintertime, bolt and flower in spring and then set seeds. Growing season is usually limited climatically with the advent of harsh season: cold winter for summer annuals and dry summer for winter annuals. Plants have a mechanism to foresee the advent of harsh season, and reproduce responding to changes in temperature and photoperiod (Salisbury, 1981). Thus, many summer annuals are short-day plants that flower when daylength becomes less than a certain threshold maximum, and most winter annuals are long-day plants that flower when daylength becomes more than a threshold minimum.

It is believed that annuals maximize their reproductive output within a growing season. To find a growth schedule that maximizes the reproductive output, simple models of plant growth have been employed. Plants are considered to consist of two parts: a vegetative body and a reproductive part (Cohen, 1971; Yokoi, 1976). The vegetative body comprises leaves, stems, and roots, and produces carbohydrates through photosynthesis. Carbohydrates are subsequently partitioned between the vegetative and the reproductive part with a control function that directs the fraction of carbohydrates allocated to vegetative or reproductive growth. Theoretical considerations have suggested that the reproductive output is maximized by a single instantaneous switch from vegetative to reproductive growth at a certain point in the growing season (Cohen, 1971; Vincent and

Pulliam, 1980; King and Roughgarden, 1982). A purely vegetative growth phase is followed by a purely reproductive growth.

Model predictions of the optimal switch have been tested for several annual species in both natural and artificial environments and found to agree well with the observed patterns (Paltridge and Denholm, 1974; Zeide, 1978; King and Roughgarden, 1983; Chiariello and Roughgarden, 1984). While actual plants showed gradual switching where the vegetative growth and reproductive growth overlap for a certain period, reduction in reproductive output due to gradual against instantaneous switch was small (King and Roughgarden, 1982).

A simple model predicts that the optimal time for the switch (t_s^*) depends on the productivity (p) of the plant and the length of the growth period (T) such that

$$t_s^* = T - \frac{1}{p} \tag{6.1}$$

(Cohen, 1971). This implies that to maximize the reproductive output, plants growing in an environment of low productivity (i.e., small p) should flower earlier than those growing in a more productive environment (i.e., large p). However, most annuals flower in response to photoperiod, indicating that plants growing, e.g., in a nutrient-poor and a nutrient-rich environment both flower simultaneously when daylengh becomes shorter/ longer than a threshold. Then a question arises how plants compromise these two different demands: plants flower in response to daylength on one hand, and maximize reproductive output on the other. Are the time of flowering and its effect on final reproductive yield different among plants growing in different environments? In the first part of this chapter, we introduce a simple model of reproductive output to describe the effect of different environments on reproductive yield. Plant growth and reproduction as influenced by nutrients and time of germination and of flowering are studied. Second, we discuss the cost of reproduction with special emphasis on nitrogen use in growth and reproduction. It is widely accepted that there are costs in reproduction (Williams, 1966; Reznick, 1985; Stearns, 1989; Obeso, 2002). However, annuals use all available resources for reproduction, causing death of the whole vegetative body by the end of the growth period to leave mature capsules (fruits). This implies that annuals reproduce at all costs. Then how can we define the cost of reproduction that is relevant in annuals? Finally, we will discuss the importance of nitrogen allocation to reproductive structures as a determinant of reproductive output.

II. Modeling of Reproductive Output

We assume a plant that consists of two parts: a vegetative body and the reproductive structure (Fig. 6.1). Only the vegetative body produces carbohydrates,

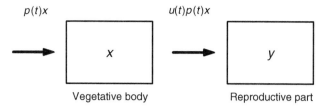

Figure 6.1   Model of growth and reproduction. The plant consists of a vegetative body (x) and a reproductive part (y). The vegetative body produces carbohydrates at a rate of p and allocated part (u) of carbohydrates to the reproductive part.

which is subsequently allocated for growth of both parts. Growth of the reproductive structure is supported by translocation of carbohydrates from the vegetative part. Let x and y be dry mass in the vegetative body and in the reproductive part, respectively. The growth of the plant is described by the following equations:

$$\frac{dx}{dt} = \left[1 - u(t)\right]p(t)x \tag{6.2}$$

$$\frac{dy}{dt} = u(t)p(t)x \tag{6.3}$$

$$x(0) > 0 \quad \text{and} \quad y(0) = 0 \tag{6.4}$$

where $p(t)$ is net dry mass productivity of the vegetative body and $u(t)$ a control function that determines the fraction of net production allocated to reproductive growth. We further assume that $u(t)$ is zero during the vegetative growth and has a constant positive value during the reproductive growth, that the growing season ends at time T, and that the plant initiates reproductive growth at $t_s (0 \le t_s \le T)$.

$$u(t) = 0 \quad \text{(for } 0 \le t < t_s) \tag{6.5a}$$

$$u(t) = u \quad \text{(for } t_s \le t \le T) \tag{6.5b}$$

From these assumptions, the reproductive ratio $[RR = y(T)/x(T)]$ was derived (Shitaka and Hirose, 1993):

if $u \ne 1$,

$$\frac{y(T)}{x(T)} = \frac{u}{1 - u}\left\{1 - \frac{1}{\exp\left[(1 - u)\int_{t_s}^{T} p(t)\,dt\right]}\right\} \tag{6.6a}$$

and if $u = 1$,

$$\frac{y(T)}{x(T)} = \int_{t_s}^{T} p(t)\,dt \tag{6.6b}$$

The optimal switching time (t_s^*), when the reproductive output is maximized for a given T, $p(t)$ and u value, is determined with $\left[\partial y(T)/\partial t_s\right]_{t_s = t_s^*} = 0$:

if $u \neq 1$,

$$\ln u + (1 - u)\int_{t_s^*}^{T} p(t)\,dt = 0 \tag{6.7a}$$

and if $u = 1$,

$$\int_{t_s^*}^{T} p(t)\,dt = 1 \tag{6.7b}$$

Equation 6.7b leads to Eq. 6.1, when $p(t) = p$ (constant). The RR for the optimal growth schedule is given by

$$\frac{y^*(T)}{x^*(T)} = u \tag{6.8}$$

If $u = 1$, it corresponds to the case of Cohen (1971).

III. Timing of Reproduction

A. Effect of Nutrient Availability

To study the effect of plant productivity on the timing of flowering and reproductive output, we grew *Xanthium canadense* Mill. plants (cocklebur, Fig. 6.2) from seeds at high and low nutrient levels (HN and LN, respectively) in a greenhouse under natural light conditions (Sugiyama and Hirose, 1991). *Xanthium* is a short-day plant, requiring at least 7.5–11 h of continuous darkness for flowering (Ray and Alexander, 1966). We observed that HN and LN plants flowered at 83 and 85 days after germination when dry mass was 14.3 and 6.2 g, respectively. The optimal time, determined by simulations of the model (above) for HN and LN plants (Fig. 6.3), was found to be 85 and 75 days after germination, respectively. We expected earlier flowering in LN than HN plants, but found only a small difference in flowering time between LN and HN, indicating that their flowering is strongly controlled by the photoperiod. A small delay in flowering in LN may be attributed to size-dependence in the response to photoperiod (Kachi and Hirose, 1983). Actual plants flowered 2 days earlier in HN and 10 days later in LN than the respective optimums. However, the effect on reproductive yield of the deviation from the optimum flowering schedule was small: the 2-day-earlier flowering in HN caused 0.2% reduction and the 10-day-later flowering in LN plants caused 2.3% reduction in the final reproductive output as compared with the theoretical optimum. Most empirical studies that tested theoretical predictions showed a good agreement between observed and predicted patterns (e.g., King and Roughgarden, 1983).

(a)

(b)

Figure 6.2 (a) Lake Kamahusa with *Xanthium canadense* Mill. population on the shore, a late summer view. (b) A close-up of *X. canadense* population. (See color plate.)

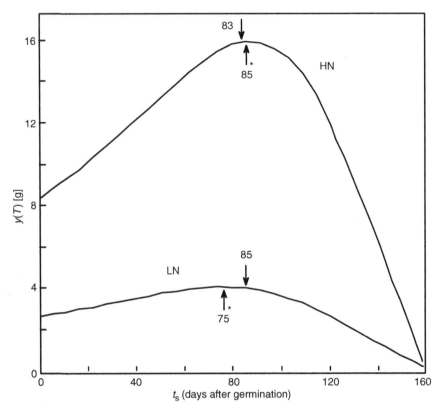

Figure 6.3 Final reproductive yield $y(T)$ as a function of the time of flowering (t_s), calculated with the model (see text). Arrows with and without an asterisk indicate optimal and actual time of flowering, respectively. *Xanthium canadense.* HN, high-nutrient plants; LN, low-nutrient plants. Redrawn with modification from Sugiyama and Hirose (1991).

Our study on *Xanthium,* however, indicated that the agreement depends on the conditions in which the plant grows. *Xanthium* followed the optimal growth schedule more closely when it grew in a high nutrient environment than when it grew in low nutrient conditions. This may not be unexpected, because *Xanthium* is known as a ruderal species that is common in disturbed areas with relatively rich nutrients and high light intensities (Grime, 1979). We may say that the photoperiodic response of *Xanthium* has been selected in such an environment.

B. Effect of Germination Dates

Seeds display polymorphism in germination rate and degree of dormancy (Harper *et al.,* 1970; Stebbins, 1971). They require safe sites for their germination and establishment (Harper, 1977; Masuda and Washitani, 1990; Eriksson and Ehrlen, 1992). Thus, the period of seed germination in the natural environment is rather broad. This is the case with *Xanthium* on the shore of a lake formed by a dam (Fig. 6.2). *Xanthium* produces a pair of somatic dimorphic seeds in their fruits (Harper *et al.*, 1970). The lower (larger) seed has no innate dormancy with a high germination potential, while the upper one is incapable of germinating at ordinary temperatures and defers germination considerably (Esashi and Leopold, 1968). The water level of the lake is high in winter through spring until the end of May; thereafter, it decreases to a minimum late in August. Many seeds that have been submerged germinate successively with retreat of the shoreline of the lake from May to July. Some seeds that have not been submerged germinate in April and the latest seedlings appear in late August. Thus, there are about four-month differences in germination time in the field.

Shitaka and Hirose (1993) examined the effect of different timing of germination on the reproductive output. It was expected that plants would initiate flowering simultaneously in late summer irrespective of different time of germination, because its flowering is strongly controlled by the photoperiod (see above). Plants germinating earlier would have a longer vegetative period than those germinating later, with the length of the reproductive period being the same as that of late germinators. We questioned the extent to which the time of flowering and consequent reproductive output deviate from the theoretical optimum. The model of reproductive output (shown above) was employed to evaluate the effect of different timing of germination.

We prepared plants that germinated on 21 May, 20 June, and 20 July and grew them in a greenhouse under natural light conditions. They flowered on 25 and 27 August, and 3 September. Delay in germination of 30 and 60 days retarded flower initiation by 2 and 9 days, respectively. Thus, the date of germination had a relatively small effect on the date of flowering. There was no significant difference in the date of plant death. All plants died early in November. The effect of germination timing on the final reproductive yield is shown in Table 6.1. Plants germinated later were smaller because of the shorter period of vegetative plant growth. However, the final

Table 6.1 Effects of Germination Time on the Vegetative Biomass [$x(T)$], Reproductive Yield [$y(T)$] at the End of the Growing Season, Reproductive Ratio [$y(T)/x(T)$], Allocation to the Reproductive Part (u), and Productivity in the Reproductive Period [$\int_{t_s}^{T} p(t)dt$] in *Xanthium canadense* (Shitaka and Hirose, 1993). Different Alphabets Within a Column Indicate a Significant Difference between Plants Germinated at Different Dates ($P < 0.05$, Tukey–Kramer test). Mean ± SE ($n = 5$)

Date of germination	$x(T)$	$y(T)$	$y(T)/x(T)$	u	$\int_{t_s}^{T} p(t)dt$
	g		No dimension		
21 May	26.0 ± 2.0[a]	7.0 ± 0.2[a]	0.27 ± 0.02[a]	0.53	0.49
20 June	16.1 ± 0.6[b]	5.9 ± 0.2[b]	0.37 ± 0.02[b]	0.59	0.64
20 July	6.3 ± 0.1[c]	4.1 ± 0.1[c]	0.64 ± 0.01[c]	0.52	1.50

reproductive yield showed less reduction than the vegetative dry mass. The RR increased 1.4 and 2.4 times with delay of 30 and 60 days in germination, respectively. Equation 6.6 implies that the RR may increase with a longer reproductive period $(T - t_s)$, a higher dry mass productivity of the vegetative part $[p(t)]$, and a higher allocation of dry mass to the reproductive part (u). Since the reproductive period and allocation of dry mass to the reproductive part (u) were similar among plants with different germination dates, different RR should be attributed to the different productivity of the vegetative body in the reproductive period $[\int_{t_s}^{T} p(t)\,dt]$. The $p(t)$ is a product of the net assimilation rate per unit leaf mass $[p_L(t)]$, and the fraction of vegetative dry mass allocated to leaf mass $[f_L(t)]$:

$$p(t) = p_L(t) \times f_L(t) \tag{6.9}$$

The higher $p(t)$ in the reproductive period of the plants germinated later resulted from a higher $p_L(t)$ and a larger $[f_L(t)]$ (Fig. 6.4).

Equation 6.7 in combination with the change in $p(t)$ (Fig. 6.4a) implies that the optimal time for switch from the vegetative to the reproductive growth should be deferred with delay in germination. However, difference in flowering time among plants with different dates of germination was small as compared to difference in optimal switch time. Plants that germinated in May and June flowered later than the optimal, whereas plants that germinated in July flowered earlier than the optimal. Since the RR is a monotonically decreasing function of t_s (Eq. 6.6; see also King and Roughgarden, 1983), RR is lower than the optimal value $(u$, see Eq. 6.8) in the plant that germinated in May and June and higher than the optimum (u) in the plants that germinated in July (Table 6.1). This may suggest that the optimal germination time is somewhere between 20 June and 20 July. In a field situation, however, plants that germinated earlier can grow taller and may outcompete late germinators through overshading and/or preemption of soil nutrients (Black and Wilkinson, 1963; Schaffer, 1977). Even though the relative allocation of dry mass to the final reproductive output was smaller in plants germinated earlier, the absolute reproductive yield was higher in early germinators. On the other hand, the longer period of the vegetative phase entails a higher risk of disturbance and herbivory before reproduction. Differences were small in the final reproductive yield divided by the length of growth period $[y(T)/T]$ among three plants: 0.042, 0.043, and 0.038 g day^{-1} in plants germinated in May, June, and July, respectively. If the mortality was higher in the plant with a longer vegetative period, the expected reproductive output per time might be smaller in early germinators.

C. Effect of Change in Flowering Time

The growth schedule of *Xanthium* is strongly constrained by the physiology of the plant: the plant flowers responding to the photoperiod. While the

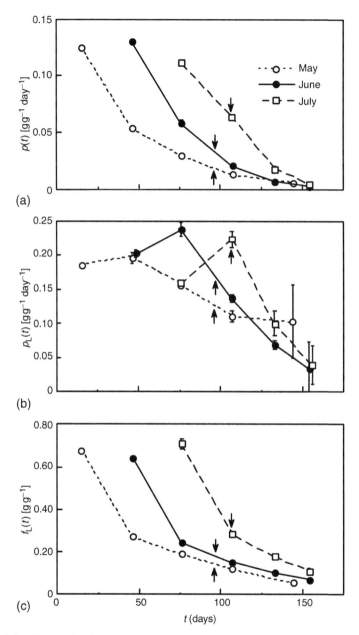

Figure 6.4 Changes in dry mass productivity per unit vegetative mass $[p(t)]$ (a), per unit leaf mass $[p_L(t)]$ (b), and the fraction of total vegetative mass in leaves $[f_L(t)]$ (c) of *Xanthium canadense* plants germinated on 21 May (open circles), 20 June (closed circles), and 20 July (open squares). Abscissa, number of days after 20 May. Vertical bars, ± SE ($n = 5$). Arrows, time of flowering. Symbols without bars indicate that the SE was less than the size of symbols. Redrawn from Shitaka and Hirose (1993).

response was within the framework of optimal growth schedule, there were systematic deviations from the theoretical prediction when plants are grown under different conditions. The model indicated that the reproductive output is determined by the productivity of the vegetative body ($p(t)$) and allocation to the reproductive part (u) in the reproductive period ($T - t_s$) (Eq. 6.6). To determine the optimal switch time, we assumed $p(t)$ that was independent of t_s. The productivity was given as a function of time, independent of flowering time. This is different from a real situation where flowering directly influences subsequent activities of the vegetative body. Particularly in annuals, flowering induces leaf senescence and finally causes destruction of the whole vegetative body to leave fruits with mature seeds (Nooden, 1988) and thus, seed production is "self-destructive" (Sinclair and de Wit, 1976). Further test of the model prediction of optimal switch is to evaluate the extent to which reproductive output changes in plants when the time of flowering is experimentally shifted. We changed the time of flowering by manipulating the photoperiod and examined the effect of change in flowering time on the subsequent growth and reproduction and the final reproductive output (Shitaka and Hirose, 1998). Manipulation of photoperiod has been successfully employed to investigate the effect of reproduction on subsequent plant growth and reproduction (Antonovics, 1980; Reekie and Bazzaz, 1987a–c, 1992). We first grew plants under noninductive photoperiod (8 h natural daylight with an additional light at midnight to interrupt continuous darkness). There were three groups of plants (EF, MF, and LF), each receiving a different period of short-day treatment (8 h natural daylight). The short-day treatment for EF (early flowering) plants started on 10 July, with MF (mid-flowering, or targeting natural flowering) plants on 14 August, and LF (late flowering) plants on 11 September. Except photoperiod treatments, plants were grown under natural light and temperature conditions. In the temperate zone, air temperature, in addition to photoperiod, changes seasonally, and delayed flowering in particular may have detrimental effects on subsequent seed maturation.

Mean dates of flower initiation were 28 July in EF, 24 August in MF, and 23 September in LF plants. The flowering date in MF plants was not different from the date of natural flowering in this area (mid to late August). In EF and MF, the vegetative part stopped growth four weeks after flowering. Reproductive structures developed rapidly after flower initiation and continued for 47 and 53 days in EF and MF plants, respectively. Transition from the vegetative to the reproductive growth was more gradual in LF plants, where the vegetative growth continued for eight weeks after flower initiation and the reproductive growth was slower and maintained longer than in the other two treatments. Although mass senescence of leaves started after flowering in all treatments, the rate of senescence was much lower in LF than in EF and MF. LF plants died due to low temperatures in winter before the process of physiological senescence was completed.

While the plant that flowered later was larger in biomass of the vegetative body [$x(T)$] due to their longer period of vegetative growth, reproductive biomass at the end of the life period [$y(T)$] was not necessarily larger in the plants that flowered later (Fig. 6.5a). Here $x(T)$ and $y(T)$ are given by

$$x(T) = \int_0^T \left[1 - u(t)\right] p(t) x \, dt \tag{6.10}$$

$$y(T) = \int_{t_s}^T u(t) p(t) x \, dt \tag{6.11}$$

The maximum reproductive output was observed in MF (targeted natural flowering) plants. The EF and LF plants had 58 and 77%, respectively, of the reproductive output of MF plants. The reproductive ratio [$y(T)/x(T)$] decreased linearly with delay in flowering. Figure 6.6 shows changes in the rate of biomass production [$p(t)x$] and its allocation to the reproductive structures [$u(t)p(t)x$]. The rate of biomass production started to decrease 3–4 weeks after flower initiation, and the reproductive allocation exceeded the biomass production later in the reproductive period. This suggests that flowering induced senescence of the whole vegetative body and that part of biomass was retranslocated from the vegetative to the reproductive part. We may define G and U such that

$$y(T) = G \times U \tag{6.12}$$

where $G = \int_{t_s}^T p(t)x \, dt$ and $U = \int_{t_s}^T u(t)p(t)x \, dt \big/ \int_{t_s}^T p(t)x \, dt$ (see Eq. 6.11). Here G is the total amount of biomass production in the reproductive period and U is the fraction of biomass allocated to the reproductive structures. G increased and U decreased with delay in flowering (Fig. 6.5b). Highest $y(T)$ in MF resulted from G being higher than EF and from U being higher than LF. Increase in G with delay in flowering was caused by a larger size of the vegetative body, whereas reduction in U was caused most probably by low air temperatures later in autumn. Mean air temperature during the reproductive growth was 21, 18, and 12°C in EF, MF, and LF plants, respectively. Low temperature might have decreased the rate of translocation of assimilates for growth and development of reproductive organs more than the rate of production of assimilates.

The present experiment showed that $x(T)/y(T)$ decreased with delay in flowering and that there was an optimal time for flowering (i.e., MF) to maximize the reproductive yield. Although this seems to be consistent with the model prediction (e.g., Cohen, 1971; King and Roughgarden, 1983), the mechanism is completely different from each other. The model assumed a fixed length of the growth period irrespective of flowering time. Plants that flowered earlier were smaller in vegetative body size, but were expected to

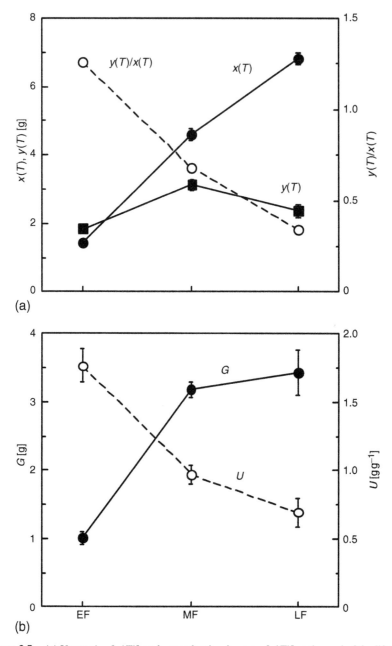

(a)

(b)

Figure 6.5   (a) Vegetative [$x(T)$] and reproductive dry mass [$y(T)$] at the end of the lifetime and the reproductive ratio [$y(T)/x(T)$]. (b) Total dry mass production (G) and the fraction allocated to the reproductive structure (U), where $y(T) = G \times U$. Mean ± SE ($n = 5$). EF, early (28 July) flowering; MF, targeted natural (24 August) flowering; LF, late (23 September) flowering plants. *Xanthium canadense*. Symbols without bars indicate that the SE was less than the size of symbols. Plotted from data in Shitaka and Hirose (1998).

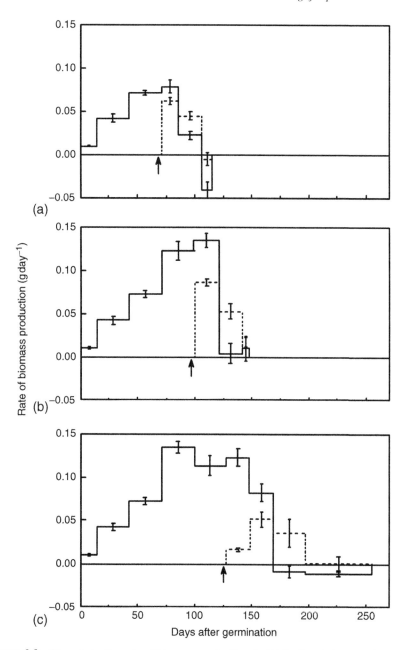

Figure 6.6 Changes in the rate of biomass production (solid line) and the growth rate of the reproductive organs (dashed line) of *Xanthium canadense* plants flowering on 28 July (a), 24 August (b), and 23 September (c). Arrows, time of flowering. Vertical bars, ± SE ($n = 5$). Redrawn from Shitaka and Hirose (1998).

use a longer period of time for reproduction. Plants flowered later were to have a shorter period for reproduction with a larger vegetative body. Thus, the model assumed that flowering did not influence the total growth period nor the assimilation rate of the vegetative body. Reproductive yield would be maximized by a compromise between the productivity (G) that would increase with body size and the decreases in the reproductive period ($T - t_s$) with delay in flowering. In reality, plants in a seasonal environment reduced the length of the growth period with an earlier flowering and extended with a later flowering. This is because reproductive growth necessarily induced senescence of leaves and the whole vegetative body and because the faster growth of reproductive structures in plants that flowered earlier shortened the length of the reproductive period. Plants flowered later retarded the senescence due to reduced rates of reproductive growth. Thus flowering strongly influenced subsequent physiological activities of the vegetative body. Real plants maximized reproductive yield by a compromise between G and U (Fig. 6.5), not by the compromise between G and ($T - t_s$), as Cohen (1971) and many others have assumed.

IV. Costs of Reproduction

We have shown that reproduction lowered the rate of biomass production and shortened the total life period by inducing senescence of the whole vegetative body. These negative effects of reproduction on subsequent growth and survival have been termed as costs of reproduction (Williams, 1966). It has been suggested that plants maximize their fitness by minimizing the cost of reproduction (Reekie and Bazzaz, 1992; Bazzaz, 1997). The cost of reproduction has been classified into two categories: direct costs of reproduction, i.e., resource allocation at the time of reproduction, and indirect costs, referred to as indirect, delayed, or demographic costs of reproduction (Bazzaz and Reekie, 1985; Obeso, 2002). Direct costs are commonly referred to as the current reproductive efforts, which are assessed by the fraction of resources allocated to reproductive structures. Indirect costs are assessed by the reduction in survival (survival costs) and in fecundity (fecundity costs), where current reproductive effort influences the probability of future survival and reproduction (Reznick, 1985, 1992; Karlsson *et al.*, 1990). However, the costs of reproduction defined in terms of losses of the potential future reproductive success may not be relevant in annuals, because annuals do not delay reproduction to the next year. Annuals, having no perennating organs, must produce seeds within a year. Nonetheless, if the cost of reproduction is due simply to the principle of allocation (Levins, 1968), i.e., when the amount of resources is limited, resources used for reproduction will not be available for growth, then this trade-off exists in annuals as well. Our experiments have shown that an increase in invest-ment in reproduction results in decreased somatic growth (Fig. 6.6). We may examine the cost of reproduction comparing reproductive output among plants flowering at different times in the growth period.

Reekie and Bazzaz (1992) controlled reproduction of long-day perennial species of *Plantago* by changing the photoperiod and examined the effect of reproduction on plant growth. The long-day treatment was expected to increase the reproductive growth at the expense of somatic growth. They defined the cost of reproduction as the reduction of biomass growth per unit increase in reproductive output: $(B_S - B_L)/(C_L - C_S)$, where B_S and B_L are the biomass of plants of long- and short-day treatment, and C_L and C_S are the capsule mass of plants of long- and short-day treatments, respec-tively. Plants of long-day treatment is reproductive while plants of short-day treatment is non- (or less) reproductive. This concept was applied to our data for *Xanthium* grown with different treatments of photoperiod to calculate the cost of reproduction in EF and MF plants against LF plants (Shitaka and Hirose, 1998): Cost of reproduction in EF = $[x_L(T_E) - x_E(T_E)]/[y_E(T_E) - y_L(T_E)] = 4.0$ g g^{-1}; cost of reproduction in MF = $[x_L(T_M) - x_M(T_M)]/[y_M(T_M) - y_L(T_M)] = 2.4$ g g^{-1}, where x and y indicate vegetative and repro-ductive biomass, respectively, T is time at the end of the life period, and suffixes E, M, and L indicate EF, MF, and LF plants, respectively. Somatic biomass decreased by 4 g to produce a unit reproductive mass in EF plants, while it decreased by 2.4 g to produce a unit reproductive mass in MF plants. Thus, the cost of reproduction was lower in natural flowering (MF) than in earlier flowering (EF). (It is impossible to calculate the cost of reproduc-tion in LF plants because no control plants were available in our experi-ment. The cost could be smaller or larger in LF.) Larger reproductive yield in natural than in earlier flowering seems to support the contention that plants maximize their fitness by minimizing the cost of reproduction. Then what caused the difference in costs of reproduction between plants with different short-day treatments?

In the rest of this chapter, we concentrate on analyzing the cost of repro-duction as an investment of dry mass and nitrogen to reproductive struc-tures (direct costs), and a reduction of vegetative growth and destruction of the somatic activities (indirect costs). We study nitrogen because its avail-ability often limits plant vegetative and reproductive growth in the natural and the agricultural environment (Chapin, 1980). Many seed crops, partic-ularly of protein seed, contain a large amount of nitrogen and the amount of nitrogen uptake from the soil in the reproductive period does not meet the high demand of nitrogen for seed growth (Sinclair and de Wit, 1975). Consequently, retranslocation of nitrogen occurs from the vegetative to the reproductive structures. This would influence the efficiency of nitrogen use for vegetative growth.

A. Reproductive Effort and the Relative Somatic Cost

The direct cost of reproduction may be evaluated by the reproductive effort (RE), which was defined by Tuomi *et al.* (1983) as:

$$RE = \frac{I_r}{(I_r + I_s)} \tag{6.13}$$

where I_r includes the amount of a specific resource allocated to reproductive parts, and I_s the somatic resource pool in reproducing plants. Tuomi *et al.* (1983) further defined the relative somatic cost of reproduction (RSC) as:

$$RSC = \frac{(I_n - I_s)}{I_n} \tag{6.14}$$

where I_n is the somatic resource pool of nonreproductive plants, and I_s the somatic resource pool of reproductive plants. RSC, being a measure of reduction in somatic growth due to reproduction, may be used to evaluate the indirect cost of reproduction. If the entire investment in reproduction (I_r) occurs at the expense of somatic growth (I_s), then $I_r + I_s = I_n$ and RE = RSC. If $I_r + I_s > I_n$, then RE > RSC, if $I_n = I_s$, then RSC is zero and if $I_n < I_s$, then RSC is negative (Thoren *et al.*, 1996). We may further add that if $I_r + I_s < I_n$, then RE < RSC. Studying the somatic cost of reproduction in three carnivorous species of *Pinguicula*, Thoren *et al.* (1996) found that RSC is always lower than RE in terms of dry mass but to a lesser extent in terms of nitrogen (Table 6.2a). Hemborg and Karlsson (1998) showed in eight subarctic species that RSC was generally similar in magnitude to the RE in terms of nitrogen and lower than the RE in terms of biomass. RSC lower than RE implies the existence of some compensation mechanisms to decrease the cost of reproduction (see also Reekie and Bazzaz, 1987a; Bazzaz, 1997). Table 6.2a shows that RE in dry mass was higher than that in nitrogen in *Pinguicula*, while RSC was not very different between dry mass and nitrogen. In a perennial grass *Agropyron repens*, Reekie and Bazzaz (1987c) also showed that RE (equivalent to their RE3) was higher than RSC (equivalent to their RE6) and that RSC was lower in terms of biomass than in terms of nitrogen.

We applied the concept of RE and RSC to the EF and MF plants of *Xanthium*, where RSC was calculated against LF plants (Table 6.2b): RE = $y_i(T_i)/[x_i(T_i) + y_i(T_i)]$ and RSC = $[x_L(T_i) - x_i(T_i)]/x_L(T_i)$, where i is either E (EF) or M (MF plants). In *Xanthium*, RE was much larger in nitrogen than in dry mass; RSC was higher than RE in dry mass, while RSC was not different from RE in nitrogen. RSC that was higher than RE in dry mass implies that reproduction caused degradation of somatic tissues without compensation (RSC > RE). Nitrogen was invested to reproduction just at

Table 6.2 Reproductive Effort (RE) and the Relative Somatic Cost of Reproduction (RSC) in Terms of Dry Mass and Nitrogen (a) for *Pinguicula alpina, P. villosa,* and *P. vulgaris* (Thoren *et al.,* 1996) and (b) for Early Flowering (EF), Targeted Natural Flowering (MF), and Late Flowering (LF) Plants of *Xanthium canadense* (calculated from Shitaka and Hirose, 1998; and unpublished data)

	RE		RSC	
Species	DM	N	DM	N
(a)				
P. alpina	0.44	0.25	0.29	0.24
P. villosa	0.72	0.48	0.26	0.27
P. vulgaris	0.52	0.32	0.28	0.19
(b)				
X. canadense				
EF	0.41	0.76	0.72	0.77
MF	0.31	0.66	0.49	0.68
LF	0.17	0.30	–	–

the expense of somatic nitrogen (RE = RSC). Thus, the pattern found in *Xanthium* was very different from that in *Pinguicula,* where RSC < RE held both in dry mass and in nitrogen (except for nitrogen in *P. alpina*). The difference could accrue from different life forms between the two genera. *Pinguicula* are perennial species that must store resources in hibernating tissues after seed production. All species that Hemborg and Karlsson (1988) studied were perennial and had RSC that was lower than RE in dry mass. RE being lower in nitrogen than in dry mass indicates that reproductive structures had lower concentrations of nitrogen than somatic tissues, and that relatively smaller amounts of nitrogen were invested for reproduction as compared to somatic growth. Growth activities of somatic tissues would not decrease or even increase with reproduction and accordingly RSC was smaller than the RE in *Pinguicula* (Thoren *et al.,* 1996). On the other hand, *Xanthium* is an annual species and all retranslocatable resources are invested to reproductive structures by the end of lifetime to leave mature capsules. Reproductive allocation (RE) of annuals is thus higher than that of perennials (Kawano, 1975; Pitelka, 1977; Primack, 1979). Annuals invest a larger fraction of plant nitrogen to reproductive structures as compared to perennials and the fraction far exceeded that of dry mass in annuals (Andel and Vera, 1977). The reproductive structures had a higher nitrogen concentration than the vegetative body (RE was higher in nitrogen than dry mass; Table 6.2), and the nitrogen cost in reproduction was larger than the dry mass cost. Retranslocation of leaf nitrogen to the reproductive structures caused detrimental effects on somatic growth through reduction in photosynthetic activities of leaves (RSC was higher than RE). RSC was maximized rather than minimized, leading to monocarpic death in annuals. Then if the cost of reproduction was evaluated by RSC, we would not say that annuals minimize the cost of reproduction to maximize the reproductive output.

B. Nitrogen Use Efficiency

Nitrogen plays a dual role in the determination of the reproductive output. More than 50% of nitrogen is involved in photosynthetic activities in leaves (Evans and Seemann, 1989; Hikosaka and Terashima, 1996) and supports dry mass production for reproductive growth on one hand, while a considerable amount of nitrogen is needed for growth of reproductive structures on the other (Sinclair and de Wit, 1975; Abrahamson and Caswell, 1982). In perennials these two roles may be met without detrimental effects on somatic growth and part of nitrogen is left for growth in the next year. In annuals like *Xanthium,* on the other hand, the reproductive structure needs a large amount of nitrogen that is not satisfied by the nitrogen uptake from soil, and somatic structures degrade to supply additional nitrogen for reproductive growth (Nooden, 1988). In natural environments plant growth is strongly constrained by the availability of nitrogen (Chapin, 1980). Nitrogen use efficiency (NUE) defined as dry mass productivity per unit nitrogen uptake has been suggested to be a useful tool to analyze the plant growth in the natural environment (Hirose, 1975; Vitousek, 1982). It was hypothesized that plants with a high NUE would be advantageous in nutrient-poor environments, but this hypothesis has not necessarily been well supported in plants growing in the natural environment. Berendse and Aerts (1987) re-defined NUE as the product of the mean residence time (MRT) and nitrogen productivity (NP; Ingestad, 1979). Plants can increase NUE by increasing the mean period during which N can be used for carbon fixation (MRT) and/or by increasing the rate of dry mass production per unit N in the plants (NP). They suggested that in a nutrient-poor environment, species with a long MRT (rather than a high NUE) have selective advantages, while in a nutrient-rich environment, species with a high NP have selective advantages. As a long MRT was often accompanied by a low NP in species growing in a nutrient-poor-environment and vice versa in a nutrient-rich environment, they further suggested that there is an evolutionary trade-off between traits leading to a long MRT and those leading to a high NP (also see Aerts and Chapin, 2000).

Eckstein and Karlsson (2001) applied the NUE as defined by Berendse and Aerts (1987) to analyzing reproduction in *Pinguicula* in a nutrient poor subarctic environment. Reproduction incurred a considerable loss of nitrogen and reduced MRT. Reduced MRT, however, was partly compensated for by an increased NP, and the effect of reproduction on NUE was relatively small. As annual plants use all available nitrogen for reproduction by the end of lifetime, we may hypothesize that timing of reproduction is determined to maximize reproductive yield per unit nitrogen uptake. We test this hypothesis in *Xanthium* plants whose flowering time was changed by manipulations of photoperiod (Shitaka and Hirose, 1998). Growth of *Xanthium* is analyzed in relation to nitrogen uptake and dry mass production.

We define NUE in plant growth at day *t* after germination:

$$NUE = \frac{\Delta W(t)}{\Delta N(t)} \tag{6.15}$$

where $\Delta W(t)$ and $\Delta N(t)$ are the cumulative amount of dry mass production and nitrogen uptake (including those in dead tissues), respectively. Nitrogen productivity (NP) at day *t* is defined by:

$$NP = \frac{1}{N}\frac{\Delta W(t)}{\Delta t} \tag{6.16}$$

where N and $\Delta W(t)/\Delta t$ are the plant nitrogen in living tissues and dry mass productivity averaged over the period Δt, respectively. As Berendse and Aerts (1987) defined NUE as the product of NP and MRT, we define MRT here as the ratio of NUE to NP:

$$MRT = \frac{NUE}{NP} \tag{6.17}$$

where NUE and NP are given by Eqs (6.15) and (6.16), respectively. Originally MRT was applied to the plant at a steady-state or the plant that was assumed to be at a steady-state. Accordingly MRT as well as NUE have been evaluated on a yearly basis. However, Eq. 6.17 gives MRT = N $\Delta t/\Delta N(t)$, i.e., MRT is defined as a function of day *t*. During plant growth, nitrogen is continuously taken up from the soil and is allocated for the vegetative and reproductive growth, in which process part of nitrogen is lost through litter production. At a certain date in plant growth, some nitrogen in the plant body had been absorbed earlier and the other later, for which situation we may calculate the mean time of residence of nitrogen by using Eq. 6.17. NP is further factorized into the product PNUE and FNL (Hirose, 1988):

$$NP = PNUE \times FNL \tag{6.18}$$

where PNUE is the photosynthetic nitrogen use efficiency and defined as:

$$PNUE = \frac{1}{N_L}\frac{\Delta W(t)}{\Delta t} \tag{6.19}$$

where N_L is the amount of nitrogen in leaves averaged over the period Δt. FNL is the fraction of plant nitrogen in leaves:

$$\text{FNL} = \frac{N_L}{N}$$

(6.20)

Figures 6.7a–e show time courses of the NUE, MRT, NP, PNUE, and FNL calculated for EF, MF, and LF plants of *Xanthium*. High NUE observed at the very beginning of plant growth (Fig. 6.7a) is attributable to the growth at this stage being less dependent on nitrogen uptake owing to dominant use of storage nitrogen in the seed. Except for this stage, NUE increased in the vegetative growth phase and decreased after flower initiation. MRT increased steadily with growth with no significant difference between plants having different flowering time (Fig. 6.7b). This is because nitrogen loss was small and most nitrogen taken up from soil was retained within the plant body until the end of lifetime. On the other hand, NP decreased sharply and flower initiation accelerated the reduction in NP (Fig. 6.7c), in which two factors were involved (Eq. 6.18): PNUE decreased through growth with smaller effects of reproduction (Fig. 6.7d), while FNL was nearly constant in the vegetative phase and decreased sharply with flower initiation (Fig. 6.7e). These results suggest that the increase in NUE in the vegetative phase was due to increase in MRT, while the reduction in NUE after flower initiation was due to reduction in NP that was caused primarily by the reduction in FNL and secondarily by the reduction in PNUE. Reduction in FNL was caused by predominant allocation of plant nitrogen to reproductive structures in the reproductive period. About 76, 66, and 30% of plant nitrogen were allocated to reproductive structures of EF, MF, and LF plants, respectively (RE in nitrogen; Table 6.2).

From the above analysis we may conclude that a large amount of nitrogen allocated to reproductive structures caused reduction in NUE of the plant (Fig. 6.7a). How are these differences in NUE translated to the difference in the final reproductive yield? Plants absorb nitrogen from soil and absorbed nitrogen is used for growth and reproduction. In annuals, all vegetative structures are destroyed by the end of lifetime to leave mature capsules (fruits). Given that capsules are the final product of an annual, the reproductive yield per unit nitrogen uptake (reproductive nitrogen use efficiency, RNUE) would be an important measure to indicate how plants used nitrogen for reproduction. Thus we define RNUE as:

$$\text{RNUE} = \text{NUE} \times \text{RE}$$

(6.21)

Figure 6.8a shows RNUE together with NUE and RE. NUE increased (see also Fig. 6.7a) and RE decreased with delay in flowering (Table 6.2). RNUE was maximized in MF (targeted natural flowering) plants as a compromise between NUE and RE. Reduction in RE with delay in flowering was

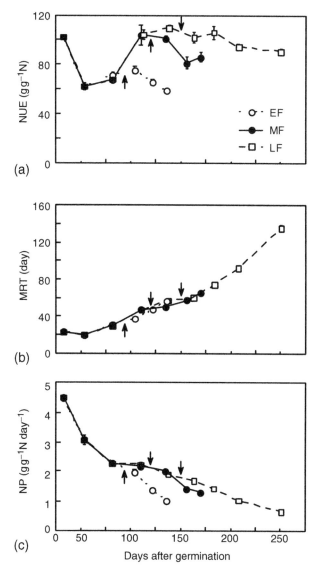

Figure 6.7 Changes in nitrogen use efficiency (NUE, a), mean residence time of nitrogen (MRT, b), nitrogen productivity (NP, c), photosynthetic nitrogen use efficiency (PNUE, d), and fraction of nitrogen in leaves (FNL, e) of early flowering (EF), targeted natural flowering (MF), and late flowering (LF) plants of *Xanthium canadense*. Mean ± SE ($n = 5$). Symbols without bars indicate that the SE was less than the size of symbols. Cumulated values are plotted against days after germination. Arrows indicate the time of flowering.

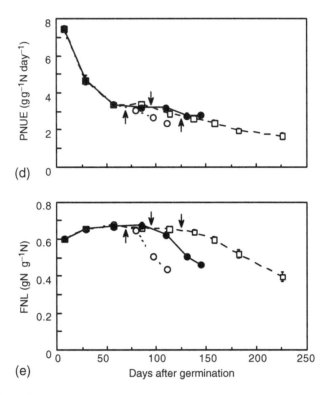

Figure 6.7, cont'd

attributed to lower temperatures, which reduced translocation of carbohydrates and nitrogen to reproductive structures (Shitaka and Hirose, 1998). Nitrogen Productivity was highest in MF, while MRT was highest in LF (Fig. 6.8b). Highest NP in MF was due to PNUE and FLN both being highest in MF plants (Fig. 6.8c). Thus, the timing of natural flowering (MF) maximized reproductive yield per unit nitrogen uptake. In the natural environment where the availability of nitrogen is limited, it would be important for plants to have a high RNUE to increase the final reproductive yield.

V. Reproductive Nitrogen

Flowering decreased or prevented a further increase in NUE of the plant (Fig. 6.7a). Figures 6.7b–e suggest that this was caused primarily by the reduction in FNL. Reproductive Effort in terms of nitrogen was high and nearly equal to RSC (Table 6.2), which indicates that the reproductive structure needs a large amount of nitrogen for growth and that nitrogen was allocated to the reproductive structure at the expense of nitrogen for

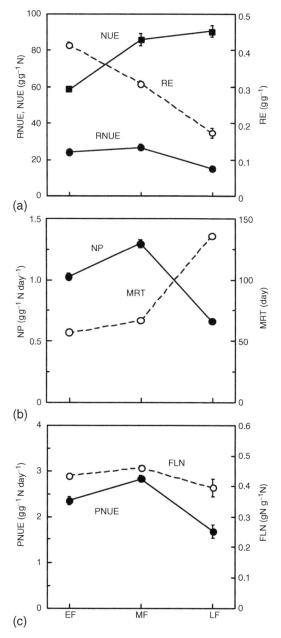

Figure 6.8 (a) Reproductive nitrogen use efficiency (RNUE), nitrogen use efficiency (NUE), and reproductive effort (RE). RNUE = NUE × RE. (b) Nitrogen productivity (NP) and mean residence time of nitrogen (MRT). NUE = NP × MRT. (c) Photosynthetic nitrogen use efficiency (PNUE) and fraction of nitrogen in leaves (FLN). NP = PNUE × FNL. Early flowering (EF), targeted natural flowering (MF), and late flowering (LF) plants of *Xanthium canadense*. Mean ± SE (*n* = 5). Symbols without bars indicate that the SE was less than the size of symbols.

vegetative growth. Reduction in nitrogen concentration in the vegetative body decreased dry mass productivity in the reproductive period, which in turn reduced the final reproductive biomass. A capsule of *Xanthium* has two dimorphic seeds in it. Seed nitrogen concentration was very high (7–9%), though the nitrogen concentration of the total reproductive structure (capsules + seeds) was lower (3–4%) due to the extremely low nitrogen concentration (0.3–0.5%) of capsule shells excluding seeds. Sinclair and de Wit (1975) suggested that in seed crops of high nitrogen concentration, nitrogen uptake in the reproductive period did not meet the high demand for seed growth and that the remaining nitrogen demand must be obtained from nitrogen in the vegetative body. They further suggested that this caused "self-destruction" of the whole plant, which limited the final reproductive yield with a shorter period of seed development.

Studying the effect of elevated CO_2 on the reproductive yield of *Xanthium*, we found that the biomass of the reproductive part increased by 53% at elevated CO_2 as compared at ambient CO_2, but that this increase was due mostly to the increase in capsule (shell) mass only, without significant effects on seed mass (Table 6.3a; Kinugasa *et al.*, 2003). There was no significant difference in the amount of nitrogen that was allocated to the reproductive structures (seeds + capsules, Table 6.3b). While CO_2 elevation decreased the nitrogen concentration of the whole reproductive part, it virtually had no effect on seed nitrogen concentration (Table 6.3c). These results suggest that seed production of *Xanthium* was not constrained by the availability of carbohydrates, but was strongly limited by the availability of nitrogen for seed growth. Kimball *et al.* (2002) reported that the boll (seed + lint) yield

Table 6.3 Reproductive Yield of *Xanthium canadense* Grown at ambient (360 µmol mol[-1]) or at Elevated (700 µmol mol[-1]) CO_2. (a) Dry Mass, (b) Nitrogen, and (c) Nitrogen Concentration (Kinugasa *et al.*, 2003). Different Alphabets Indicate a Significant Difference between Ambient and Elevated CO_2 ($P < 0.05$, ANOVA). Mean ± SE ($n = 10$)

CO_2	Reproductive part (seeds + capsules)	Seeds	Capsules
(a) Dry mass (g)			
Ambient	3.79 ± 0.25[a]	1.51 ± 0.13[a]	2.28 ± 0.26[a]
Elevated	5.80 ± 0.49[b]	1.56 ± 0.32[a]	4.24 ± 0.39[b]
(b) Nitrogen (g)			
Ambient	0.149 ± 0.010[a]	0.136 ± 0.011[a]	0.013 ± 0.003[a]
Elevated	0.163 ± 0.017[a]	0.127 ± 0.024[a]	0.036 ± 0.009[b]
(c) Nitrogen concentration (%)			
Ambient	3.99 ± 0.25[a]	9.11 ± 0.17[a]	0.51 ± 0.06[a]
Elevated	2.81 ± 0.19[b]	8.55 ± 0.33[a]	0.83 ± 0.17[a]

of cotton increased 40% by free air CO_2 enrichment, while lint fiber portion of the yield increased even more, by 54%. Carbon dioxide elevation seems to increase the biomass of tissues with a low nitrogen concentration to a larger extent than those with high nitrogen.

We may hypothesize that the reproductive yield is determined by the availability of nitrogen for seed growth, particularly in plants whose seeds have a high nitrogen concentration. If the amount of nitrogen uptake from soil does not satisfy the demand for reproductive growth, the rest must be supplied from retranslocation of nitrogen in the vegetative body, causing senescence of leaves and the whole vegetative body in annuals. This incurs the costs of reproduction. Now suppose a plant that has α g of nitrogen in the vegetative body at the beginning of flowering, that δ of the plant nitrogen is retranslocatable for reproductive growth ($0 < \delta < 1$), and that during the reproductive period, the plant continues the uptake of nitrogen from soil at a rate of β g day^{-1} (Fig. 6.9a). If the reproductive structure having a nitrogen concentration of n develops at a constant rate of γ g day^{-1}, the reproductive structure demands γnt g of nitrogen by time t after flower initiation. Then,

$$\text{N demand} = \gamma nt \tag{6.22}$$

This nitrogen must be supplied from nitrogen uptake from soil (βt) or degradation of nitrogen compounds in the vegetative body ($\alpha\delta$).

$$\text{N supply} = \alpha\delta + \beta t \tag{6.23}$$

We may assume that the reproductive structure stops growth when nitrogen demand exceeds supply:

$$\gamma nt = \alpha\delta + \beta t \tag{6.24}$$

This equation defines the length of the reproductive period (t):

$$t = \frac{\alpha\delta}{\gamma n - \beta} \tag{6.25}$$

As the amount of dry mass that is allocated to the reproductive structure is given by γt, the final reproductive yield (y) is found:

$$y = \frac{\alpha\delta\gamma}{\gamma n - \beta} \tag{6.26}$$

If plant nitrogen decreases linearly with reproductive growth, mean plant nitrogen is given by $\alpha(1 - \delta/2)$. Then, the nitrogen productivity (NP) in the reproductive period is found:

$$\text{NP} = \frac{\gamma}{\alpha(1 - \delta/2)} \tag{6.27}$$

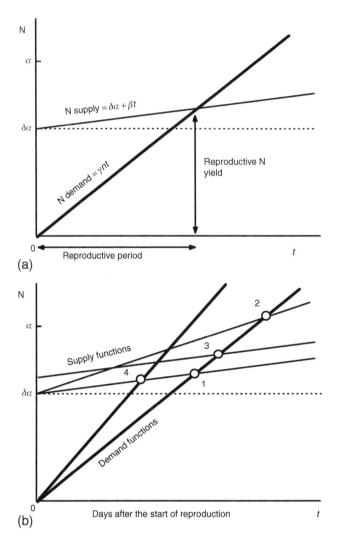

Figure 6.9   Model of reproductive yield. (a) Nitrogen for reproductive growth is supplied from new absorption of nitrogen from soil and retranslocation of nitrogen in the vegetative body. N demand = γnt; N supply = $\alpha\delta + \beta t$, where t is days after the start of reproduction. The reproductive structure having a constant nitrogen concentration (n) grows at a constant rate (γ). The plant absorbs nitrogen at a constant rate (β). Growth of the reproductive part stops when the demand exceeds the supply. Reproductive yield is given by $\alpha\delta\gamma/(\gamma n - \beta)$. (b) Effects of change in demand (thick line) and supply function (thin line) on the length of the reproductive period and final reproductive yield. Reproductive period and reproductive N yield would change from 1 to 2 when the uptake rate (β) was increased; to 3 when the resorption rate (δ) was higher; or to 4 when the growth rate (γ) was increased or when plants had a higher seed N concentration (n).

Thus, NP is independent of nitrogen concentration of the reproductive structure (n), indicating that there is a one-to-one correspondence between γ and NP for a given α and δ. We may assume a constant NP instead of a constant growth rate.

Equation 6.26 implies that reproductive yield increases with increasing supply of nitrogen [represented by the rate of nitrogen uptake (β) and the amount of retranslocatable nitrogen in the vegetative body ($\alpha\delta$)], and decreases with increasing demand of reproductive nitrogen (represented by γ and n) (see Fig. 6.9b). If we substitute $\alpha = 70$ mg N, $\beta = 1$ mg N day^{-1}, $\gamma = 60$ mg day^{-1}, $\delta = 0.48$, and $n = 0.025$ (approximate values observed in MF, i.e., targeted natural flowering plants), then the final reproductive mass is calculated as 4.03 g and the reproductive period is 67 days. Deviations from the observed values (3.14 g and 52 days, respectively) should be accounted for by parameter values of actual plants being not constant in the reproductive period. NP is calculated as 1.06 g g^{-1}N day^{-1}, the value within the range observed in the reproductive period (Fig. 6.7c). If the plant had $n = 0.02$ (or 0.03) with the other parameters being kept constant, the reproductive yield would be 10.1 (or 2.52) g with the reproductive period of 168 (or 42) days. Thus, the reproductive yield and the length of the reproductive period are very sensitive to the nitrogen concentration of reproductive structures.

High nitrogen demand for reproductive growth reduces the length of reproductive period as well as the reproductive yield (Fig. 6.9b). Using a simulation model for soybean growth, Sinclair and de Wit (1976) demonstrated that self-destructive characteristics of soybean accompanying seed-filling limited the length of the seed development period and thereby placed a potential limit on total seed production. Even in rice whose grain has a lower nitrogen concentration, only 10–30% of total reproductive nitrogen was supplied from newly absorbed nitrogen and the rest was remobilized from nitrogen in vegetative tissues, which accelerated leaf senescence (Mae, 1997). A monocarpic biennial *Arctium tomentosum,* producing seeds with high nitrogen concentration, had a high harvest index for nitrogen (0.73) as compared to that for dry mass (0.19) (Heilmeier *et al.,* 1986). Monocarpic species (annuals and biennials) are self-destructive and have a high δ for reproductive growth to maximize reproductive yield. On the other hand, perennials, leaving part of nitrogen for growth in the next year, may have a lower δ for reproductive growth. A perennial herb *Solidago altissima* allocated 1.1 g N m^{-2} stand area for sexual reproduction, the amount of which was much smaller than that (4.7 g N m^{-2}) for storage in rhizomes for next year's growth (Hirose, 1971). Nitrogen concentration of reproductive structures of some perennials was not high and their RE in terms of nitrogen was lower than that in dry mass (Table 6.2a; Thoren *et al.,* 1996).

VI. Conclusions

Availability of nitrogen strongly limits the reproductive output in annuals, particularly in those who have seeds of a high nitrogen concentration. Annuals reproduce within a year at all costs and the costs are represented by translocation of nitrogen from the vegetative to the reproductive structures. Reproductive yield is determined by the amount of nitrogen allocated for reproductive growth and nitrogen concentration of reproductive structures. Natural flowering time maximized the reproductive output by maximizing the reproductive nitrogen use efficiency (RNUE), reproductive output per unit amount of nitrogen uptake from soil.

References

Abrahamson G. W. and Caswell H. (1982) On the comparative allocation of biomass, energy, and nutrients in plants. *Ecology* **63**: 982–991.

Aerts R. and Chapin F. S. III. (2000) The mineral nutrition of wild plants revisited: A re-evaluation of processes and patterns. *Advances in Ecological Research* **30**: 1–67.

Andel J. van and Vera F. (1977) Reproductive allocation in *Senecio sylvaticus* and *Chamaenerion angustifolium* in relation to mineral nutrition. *Journal of Ecology* **65**: 747–758.

Antonovics J. (1980) Concepts of resource allocation and partitioning in plants. In *Limits to Action* (J. E. R. Staddon, ed.) pp. 1–35, Academic Press, New York.

Bazzaz F. A. (1997) Allocation of resources in plants: state of the science and critical questions. In *Plant Resource Allocation* (F. A. Bazzaz and J. Grace, eds.), pp. 1–37, Academic Press, San Diego.

Bazzaz F. A. and Reekie E. (1985) The meaning and measurement of reproductive effort in plants. In *Studies on Plant Demography: A Festschrift for John L. Harper* (J. White, ed.), pp. 373–387, Academic Press, London.

Berendse F. and Aerts R. (1987) Nitrogen-use-efficiency: a biologically meaningful definition? *Functional Ecology* **1**: 293–296.

Black J. N. and Wilkinson G. N. (1963) The role of time of emergence in determining the growth of individual plants in swards of subterranean clover (*Trifolium subterraneum* L.). *Australian Journal of Agricultural Research* **14**: 628–638.

Chapin F. S. III. (1980) The mineral nutrition of wild plants. *Annual Review of Ecology and Systematics* **11**: 233–260.

Chiariello N. and Roughgarden J. (1984) Storage allocation in seasonal races of an annual plant: optimal versus actual allocation. *Ecology* **65**: 1290–1301.

Cohen D. (1971) Maximizing final yield when growth is limited by time or by limiting resources. *Journal of Theoretical Biology* **33**: 299–307.

Eckstein R. L. and Karlsson P. S. (2001) The effect of reproduction on nitrogen use-efficiency of three species of the carnivorous genus *Pinguicula*. *Journal of Ecology* **89**: 798–806.

Eriksson O. and Ehrlen J. (1992) Seed and microsite limitation of recruitment in plant populations. *Oecologia* **91**: 360–364.

Esashi Y. and Leopold A. C. (1968) Physical forces in dormancy and germination of *Xanthium* seeds. *Plant Physiology* **43**: 360–364.

Evans J. R. and Seemann J. R. (1989) The allocation of protein nitrogen in the photosynthetic apparatus: costs, consequences and control. In *photosynthesis* (W. R. Brigs, ed.), pp. 183–205, Alan R. Liss, New York.

Grime J. P. (1979) *Plant Strategies and Vegetation Processes.* John Wiley, Chichester.

Harper J. L. (1977) *Population Biology of Plants*. Academic Press, New York.

Harper J. L., Lovell P. H. and Moore K. G. (1970) The shapes and sizes of seeds. *Annual Review of Ecology and Systematics* 1: 327–356.

Heilmeier H., Schultze E. D. and Whale D. M. (1986) Carbon and nitrogen partitioning in the biennial monocarp *Arctium tomentosum* Mill. *Oecologia* 70: 466–474.

Hemborg A. M. and Karlsson P. S. (1998) Somatic costs of reproduction in eight subarctic plant species. *Oikos* 82: 149–157.

Hikosaka K. and Terashima I. (1996) Nitrogen partitioning among photosynthetic components and its consequence in sun and shade plants. *Functional Ecology* 10: 335–343.

Hirose T. (1971) Nitrogen turnover and dry-matter production of a *Solidago altissima* population. *Japanese Journal of Ecology* 21: 18–32.

Hirose T. (1975) Relations between turnover rate, resource utility and structure of some plant populations: A study in the matter budgets. *Journal of the Faculty of Science,* The University of Tokyo, Section III, 11: 355–407.

Hirose T. (1988) Modelling the relative growth rate as a function of plant nitrogen concentration. *Physiologia Plantarum* 72: 185–189.

Ingestad T. (1979) Nitrogen stress in birch seedlings. II. N, P, Ca and Mg nutrition. *Physiologia Plantarum* 45: 149–157.

Kachi N. and Hirose T. (1983) Bolting induction in *Oenothera erythrosepala* Borbas in relation to rosette size, vernalization, and photoperiod. *Oecologia* 60: 6–9.

Karlsson P. S., Svensson B. M., Carlsson B. A. and Nordell K. O. (1990) Resource investment in reproduction and its consequences in three *Pinguicula* species. *Oikos* 59: 393–398.

Kawano S. (1975) The productive and reproductive biology of flowering plants. II. The concept of life history strategy in plants. *Journal of the College Liberal Arts of Toyama University, Japan* 8: 51–86.

Kimball B. A., Kobayashi K. and Bindi M. (2002) Responses of agricultural crops to free-air CO_2 enrichment. *Advances in Agronomy* 77: 293–368.

King D. and Roughgarden J. (1982) Multiple switches between vegetative and reproductive growth in annual plants. *Theoretical Population Biology* 21: 194–204.

King D. and Roughgarden J. (1983) Energy allocation patterns of the California grassland annuals *Plantago erecta* and *Clarkia rubicunda*. *Ecology* 64: 16–24.

Kinugasa T., Hikosaka K. and Hirose T. (2003) Reproductive allocation of an annual *Xanthium canadense* growing in elevated CO_2. *Oecologia* 137: 1–9.

Levins R. (1968) *Evolution in Changing Environment*. Princeton University Press, Princeton, New Jersey.

Mae T. (1997) Physiological nitrogen efficiency in rice: nitrogen utilization, photosynthesis, and yield potential. In *Plant Nutrition - For Sustainable Food Production and Environment* (T. Anto, ed.), pp. 51–60, Kluwer Academic Publishers, Dordrecht.

Masuda M. and Washitani I. (1990) A comparative ecology of the seasonal schedules for 'reproduction by seeds' in a moist tall grassland community. *Functional Ecology* 4: 169–182.

Nooden L. D. (1988) Whole plant senescence. In *Senescence and Aging in Plants* (L. D. Nooden and A. C. Leopold, eds.), pp. 391–439, Academic Press, London.

Obeso J. R. (2002) The costs of reproduction in plants. *New Phytologist* 155: 321–348.

Paltridge G. W. and Denholm J. V. (1974) Plant yield and the switch from vegetative to reproductive growth. *Journal of Theoretical Biology* 44: 23–34.

Pitelka L. F. (1977) Energy allocation in annual and perennial lupines (*Lupinus*: Leguminosae). *Ecology* 58: 1055–1065.

Primack R. B. (1979). Reproductive effort in annual and perennial species of *Plantago* (Plantaginaceae). *American Naturalist* 114: 51–62.

Ray P. M. and Alexander W. E. (1966) Photoperiodic adaptation to latitude in *Xanthium strumarium*. *American Journal of Botany* 53: 806–816.

Reekie E. G. and Bazzaz F. A. (1987a) Reproductive effort in plants. 1. Carbon allocation to reproduction. *American Naturalist* **129**: 876–896.

Reekie E. G. and Bazzaz F. A. (1987b). Reproductive effort in plants. 2. Does carbon reflect the allocation of other resources? *American Naturalist* **129**: 897–906.

Reekie E. G. and Bazzaz F. A. (1987c) Reproductive effort in plants. 3. Effect of reproduction on vegetative activity. *American Naturalist* **129**: 907–919.

Reekie E. G. and Bazzaz F. A. (1992) Cost of reproduction as reduced growth in genotypes of two congeneric species with contrasting life histories. *Oecologia* **90**: 21–26.

Reznick D. (1985) Cost of reproduction: an evaluation of the empirical evidence. *Oikos* **44**: 257–267.

Reznick D. (1992) Measuring the costs of reproduction. *Trends in Ecology and Evolution* **7**: 42–49.

Salisbury F. B. (1981) Responses to photoperiod. In *Physiological Plant Ecology. I. Encyclopedia of Plant Physiology, New Series Vol 12A* (O. L. Lange, P. S. Nobel, C. B. Osmond, H. Ziegler, eds.), pp. 135–167, Springer, Berlin.

Schaffer W. M. (1977) Some observations on the evolution of reproductive rate and competitive ability in flowering plants. *Theoretical Population Biology* **11**: 90–104.

Shitaka Y. and Hirose T. (1993) Timing of seed germination and the reproductive effort in *Xanthium canadense*. *Oecologia* **95**: 334–339.

Shitaka Y. and Hirose T. (1998) Effects of shift in flowering time on the reproductive output of *Xanthium canadense* in a seasonal environment. *Oecologia* **114**: 361–367.

Sinclair T. R. and de Wit C. T. (1975) Photosynthate and nitrogen requirements for seed production by various crops. *Science* **189**: 565–567.

Sinclair T. R. and de Wit C. T. (1976) Analysis of the carbon and nitrogen limitations to soybean yield. *Agronomy Journal* **68**: 319–324.

Stearns S. C. (1989) Trade-offs in life-history evolution. *Functional Ecology* **3**: 259–268.

Stebbins G. L. (1971) Adaptive radiation of reproductive characteristics in angiosperms, II: seeds and seedlings. *Annual Review of Ecology and Systematics* **2**: 237–260.

Sugiyama H. and Hirose T. (1991) Growth schedule of *Xanthium canadense:* Does it optimize the timing of reproduction? *Oecologia* **88**: 55–60.

Thoren L. M., Karlsson P. S. and Tuomi J. (1996) Somatic cost of reproduction in three carnivorous *Pinguicula* species. *Oikos* **76**: 427–434.

Tuomi J., Hakala T. and Haukioja E. (1983) Alternative concept of reproductive efforts, costs of reproduction and selection in life-history evolution. *American Zoologist* **23**: 25–34.

Vincent T. L. and Pulliam H. R. (1980) Evolution of life history strategies for an asexual annual plant model. *Theoretical Population Biology* **17**: 215–231.

Vitousek P. M. (1982) Nutrient cycling and nutrient use efficiency. *American Naturalist* **119**: 553–572.

Williams G. C. (1966) *Adaptation and Natural Selection.* Princeton University Press, Princeton, New Jersey.

Yokoi Y. (1976) Growth and reproduction in higher plants. I. Theoretical analysis by mathematical models. *Botanical Magazine Tokyo* **89**: 1–14.

Zeide B. (1978) Reproductive behavior of plants in time. *American Naturalist* **112**: 636–639.

7

The Shape of the Trade-off Function between Reproduction and Growth

Edward G. Reekie, German Avila-Sakar

I. Introduction

Implicit in virtually all the various life history theories and models that have been proposed to explain variation in reproductive allocation, is the assumption that allocating resources to reproduction results in a "cost" to the plant (Williams, 1966; Stearns, 1989; Willson, 1983). This cost should be evident in the components of fitness; i.e., increased allocation to current reproduction should decrease future reproduction. In a nonclonal plant, this essentially means that current reproduction should either decrease the likelihood of surviving to the next growing season(s), or the number of propagules produced in future seasons. In a clonal plant, the situation is a little more complicated in that "reproduction" can also occur by vegetative propagation. In this case, one must also consider the cost in terms of the number of vegetative propagules produced. Regardless of whether you are considering a clonal or nonclonal plant, however, the mechanistic link between current and future reproduction is vegetative growth (Bazzaz *et al.*, 2000). Allocating resources to current reproduction deprives vegetative growth of these resources and so reduces plant size and resource storage. Reduction in growth in turn leads to reduced vegetative propagation, future survival, and future reproduction. There is abundant empirical evidence that vegetative growth is indeed closely linked to all the various measures of plant fitness (e.g., Farris and Lechowicz, 1990; Mitchell-Olds and Bergelson, 1990; Schwaegerle and Levin, 1990).

Although there is a general agreement that there is indeed a trade-off between current and future reproduction in plants that is mediated through the trade-off between reproduction and growth, there is very little empirical data available on what this trade-off function actually looks like.

Most researchers, either implicitly or explicitly assume that it is a linear function such that an increase in current reproduction results in a proportional decrease in growth (Fig. 7.1a). It is possible however, to develop hypotheses that predict curvilinear trade-off functions. A concave curvilinear function (Fig. 7.1b) would result if there were "economies of scale" associated with increasing the level of current reproduction to a higher level. One example of such an "economy" would be if a plant increased reproductive output by increasing the number of flowers in an inflorescence; since the stalk required to support the inflorescence was already constructed, increasing the number of flowers would be at relatively little cost. It is also possible to devise plausible hypotheses that predict a convex trade-off function (Fig. 7.1c) as discussed below. Whether the trade-off function is linear, concave, or convex has obvious implications for the predictions of life history theory (Calvo and Horvitz, 1990; Schmid, 1990). Providing the initial slopes of the trade-off functions are the same, a convex trade-off function will result in selection for relatively low levels of current reproduction relative to that expected for a linear trade-off function. On the other hand, a concave trade-off function will result in selection for relatively high levels of reproductive output compared to a linear function with the same initial slope.

In this chapter, we summarize recent research from our laboratory using a model plant species, *Plantago major*, which addresses the question of what this trade-off function actually looks like. It is not intended to be a review of the literature on this topic. Rather, it represents our particular perspective on this issue and how this perspective has evolved.

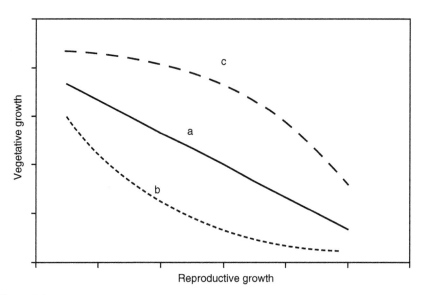

Figure 7.1   Three hypothetical trade-off functions between reproduction and vegetative growth.

II. Methods of Describing the Trade-off Function

Given that the shape of the trade-off function between current and future reproduction is of fundamental importance in understanding the evolution of different life history patterns, why is there so little empirical data on what this function actually looks like? The answer lies in the formidable difficulties associated with trying to realistically assess the costs associated with reproduction (Reznick, 1985; Stearns, 1989; Bailey, 1992). Four basic approaches have been used to try to assess reproductive costs in plants. All four of these approaches have inherent problems that limit their usefulness.

The first approach is to examine the phenotypic correlation between reproduction and subsequent performance (e.g., Law, 1979; Piñero *et al.*, 1982); you measure reproductive output of individuals in year 1 and then compare how individuals that differed in reproductive output in year 1 perform in subsequent years. This deceptively simple approach is unlikely to provide an accurate measure of the cost of reproduction due to the fact that natural variation in reproductive output is likely confounded with environmental variation that will affect subsequent performance (Fox and Stevens, 1991). Those individuals who had a relatively high reproductive output in year 1 may well be growing in particularly favorable microsites, while those individuals with a low reproductive output may be growing in unfavorable microsites. Therefore, differences in the subsequent performance of these individuals may simply be a reflection of the differences in habitat rather than differences in reproductive output.

The second approach partitions phenotypic correlations into environmental versus genetic components and uses the genetic correlation between current and future reproduction to quantify the cost of reproduction (e.g., Geber, 1990). This is a labor-intensive approach, in that a large number of genetic families (clones, half-sibs, or full-sibs) that differ in reproductive output must be generated. However, it has a major advantage over the previous approach as environmental variation is no longer confounded with differences in reproductive output. Further, the measure of the reproductive cost that is obtained is directly relevant to evolutionary questions, since it is based on genetic variation that can be acted upon directly by selection (Reznick, 1985). The major drawback of this approach is finding sufficient genetic variation in reproductive output to assess reproductive costs. Given that traits closely related to fitness will be subject to strong selection pressure, there may be relatively little genetic variation in reproductive allocation within many populations (Bailey, 1992).

The third approach circumvents this problem by using artificial selection to generate genetic lines that differ in allocation to current reproduction and then assesses its effect on subsequent performance (Reznick, 1985; Stearns, 1989). This approach has been most commonly applied to animal species with short generation times. Given the length of their life cycles, this is a time-consuming procedure for most plant species. Given that many crop species have been subject to selection for increased reproductive allocation, it is possible to draw inferences regarding reproductive costs by comparing different crop selections. Unfortunately, these selections usually differ in other traits not directly related to reproductive allocation (e.g., palatability, disease, and pest resistance, etc.), making these comparisons somewhat problematic.

The fourth approach to assessing the cost of reproduction is to manipulate level of reproduction experimentally and compare the subsequent performance of the different treatments (e.g., Reekie and Bazzaz, 1992; Reekie and Reekie, 1991; Thoren *et al.*, 1996). Level of reproduction can be controlled by various means including: photoperiod manipulations in photoperiod sensitive species, growth regulator applications, removal of developing reproductive structures, and by preventing or supplementing natural pollination levels. This approach has several advantages over the previous methods. As reproduction is controlled experimentally, different levels of reproduction are not confounded with random environmental variation and one is not limited by the available genetic variation within the population. Further, these experiments can be conducted over relatively short time spans. This method has two major disadvantages however. First, the manipulations used to control the level of reproduction may introduce artifacts. For example, photoperiod manipulations may affect other aspects of physiology in addition to the induction of reproduction. Second, phenotypic manipulations of the level of reproduction may not realistically simulate genetic differences in reproductive allocation. If we are interested in explaining the evolution of different life history patterns, we must assume phenotypic and genetic differences in reproductive allocation have similar effects. This may be a reasonable assumption, but there is no easy way to rigorously test it.

Given the problems associated with accurately measuring reproductive costs discussed earlier, it is perhaps not surprising there has been relatively little effort devoted to quantitatively describing the shape of the trade-off function between current and future reproduction. Rather, efforts have been focussed on simply documenting that there is a cost associated with reproduction. These empirical studies have produced mixed results. Although there are many studies that describe negative effects of reproduction on future performance (Kozlowski, 1971; Piñero *et al.*, 1982; Primack and Hall, 1990; Newell, 1991; Cipollini and Whigham, 1994), there are many others that fail to detect any negative effects (Horvitz and Schemske, 1988; Karlsson *et al.*, 1990; Jennersten, 1991; Pfister, 1992; Ramsey, 1997), or highly variable effects depending upon environmental conditions (Zimmerman, 1991; Syrjanen and Lehtilä, 1993; Ågren and Willson, 1994; Ramadan *et al.*, 1994; Saikkonen *et al.*, 1998; Primack *et al.*, 1994). The failure to find a cost associated with reproduction has been largely attributed to the problems associated with measuring these costs rather than a true absence of cost (Reznick, 1985; Stearns, 1989; Bailey, 1992). It is, after all, difficult to accept that reproduction could occur without cost as the resources invested in reproductive structures can be very substantial; it seems only logical to assume this would have negative consequences on future reproduction. With this in mind, we conducted a study to determine which of the various methods commonly used to assess the cost of reproduction in plants has the greatest power to detect these costs (Reekie, 1998b).

Plantago major was used as the study species. This species has several advantages for studies of reproductive allocation; it is a nonclonal perennial with a fairly simple morphology, and clear separation between vegetative and reproductive structures (Hawthorn, 1974). In the vegetative state, it consists of a basal rosette of leaves attached to a caudex with a fibrous root system. Reproduction can be controlled experimentally by photoperiod manipulations as it is a long-day plant with a critical photoperiod of 14 hours. The reproductive structures are spikes borne on erect, leafless scapes. The reproductive structures contain chlorophyll and potentially supply a portion of their own carbon needs through photosynthesis.

A preliminary experiment was conducted to screen genetic families for differences in the extent of reproductive allocation. Individuals were isolated from 42 different sites representing the full range of habitats occupied by this species within the local region (Reekie, 1998a). A subset of 15 half-sib families representing the full range of available genetic variation in reproductive allocation were then grown under controlled environment conditions in which half the individuals in each family were induced to reproduce, while the remaining half were maintained in the vegetative state through photoperiod manipulations. As very low levels of light ($8 \ \mu\text{mol m}^{-2}\text{s}^{-1}$) were used to extend the photoperiod for the reproductive plants, the total amount of photosynthetically active radiation received by the two groups was approximately the same. In the fall of the first year of this study (i.e., after reproductive induction but before seed maturation), plants were transplanted to one of the two field sites (mown versus unmown turf). Reproductive output, survivorship, and growth were assessed for each individual over the next 18 months under these conditions. Reproductive costs were estimated in one of three ways: (1) the phenotypic and, (2) genetic correlations between reproduction in the first year and subsequent performance among those individuals that were induced to reproduce in the first year, and, (3) the difference in subsequent performance between those individuals that were vegetative in the first year and those that were induced to reproduce in the first year. Although evidence of reproductive costs was detected with all three methods, it was the third method (i.e., experimental manipulation of reproductive output) that provided the clearest indication that there were significant costs associated with reproduction (Table 7.1).

Table 7.1 A Comparison of Three Methods of Assessing Reproductive Costs in *Plantago major* (See Text for Details). The Effect of Differences in Reproduction among Individuals in Year 1 upon Subsequent Performance over an 18 Month Period was Examined in an Experimental Field Study in Which Individuals Were Transplanted into Either Mown or Unmown Grass Turf (Reekie, 1998b). Plant Size was Assessed at Intervals over This 18 Month Period and Capsule Mass in the Unmown Plots was Determined in the Second Year. A Dash Indicates That No Significant Cost was Detected, While Significant Costs are Indicated With an X

Measure of performance	Environment	Method of Assessing, Reproductive Costs		
		Phenotypic correlations	Genetic correlations	Experimental manipulation
Plant size at time 1	Unmown grass	X	–	X
Plant size at time 2		X	X	X
Plant size at time 3		–	X	X
Capsule mass in year 2		–	–	X
Plant size at time 1	Mown grass	–	–	X
Plant size at time 2		–	–	X
Plant size at time 3		–	–	X

Costs detected with the first two methods were transitory and were not observed for all measures of subsequent performance, whereas experimental manipulations detected highly significant costs under all conditions and for all measures of subsequent performance.

Although the correlations between current reproduction and subsequent performance provided only weak evidence for reproductive costs, they do offer one advantage over the experimental manipulation used in this particular experiment as they provided a range of different levels of reproductive investment. The photoperiod manipulation provided only two levels of reproductive investment (vegetative versus reproductive individuals) therefore, it was not possible to assess the shape of the trade-off function with the manipulations. However, the genetic correlations provide some evidence that this trade-off function was nonlinear and convex (Fig. 7.2). This evidence is weak since there was a great deal of scatter in the data and the relationship was only marginally significant, but it is suggestive.

Given that experimental manipulations provide much greater control over the level of reproduction and that the level of reproduction is not confounded with random environmental or genetic variation, it is not surprising that they provide the most sensitive measure of reproductive costs. Therefore, provided one is willing to accept the assumption that phenotypic manipulations provide a measure of cost that is approximately equivalent to costs assessed at the level of genetic differences, experimental

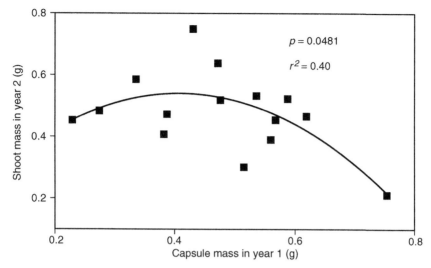

Figure 7.2 Effect of reproduction on subsequent performance in *P. major* growing in a grass sward (Reekie, 1998b). Capsule mass in 1990 was used as the measure of reproductive investment and shoot mass in May of 1992 was used as the measure of subsequent performance. Individual points represent a mean of nine individuals for each of 15 half-sibling families. The 15 families were chosen on the basis of a preliminary experiment with a larger sample of families to represent the full range of genetic variation in reproductive investment present in this species. The line was fitted by means of least squares regression and represents the genetic trade-off between reproduction and future performance.

manipulations provide the best hope of assessing the shape of the trade-off function, if they can be used to induce a wide range of different levels of reproductive investment. In the next section, we describe the results of such a study (Reekie *et al.*, 2002).

III. The Shape of the Trade-off Function in *Plantago*

Plantago major and *P. rugelii* were used as the study species. *Plantago rugelii* is morphologically very similar to *P. major*, but is found on less disturbed sites, delays reproduction to an older age and allocates a smaller proportion of its biomass to reproductive structures (Hawthorn and Cavers, 1976). Individuals of both species were grown under controlled environment conditions. Individuals were given one of nine photoperiod treatments: 0, 1, 2, 3, 4, 5, 6, 7, or 8 weeks of an inductive photoperiod (14 h). Over the remaining time in the 13 week growth period, plants received a noninductive photoperiod (12 h). As low levels of light ($15\,\mu mol\,m^{-2}\,s^{-1}$) were used to extend the noninductive photoperiod ($600\,\mu mol\,m^{-2}\,s^{-1}$), the total amount of light received by the different photoperiod treatments differed by less than 0.5%. All plants were harvested at the end of the 13 week growth period and the biomass of vegetative and reproductive structures determined.

The photoperiod treatments successfully induced a wide range of reproductive output in both species, but the level of reproductive output was lower in *P. rugelii* (Fig. 7.3a,c). In both species, the increase in reproductive output with the number of inductive weeks was approximately linear with the slope of this increase being steeper in the case of *P. major*. There was little evidence that the number of inductive weeks (i.e., reproduction) had any negative effect upon vegetative growth in *P. rugelii* (Fig. 7.3d). In *P. major*, there was a sharp decline in vegetative growth once the number of inductive weeks exceeded two, but no evidence of any negative effects at either one or two inductive weeks (Fig. 7.3b).

Taking the data of the two species together and plotting vegetative biomass as a function of reproductive biomass, suggests that the trade-off function in *Plantago* is convex with little evidence of any negative effect of reproduction on growth below a reproductive biomass of 1.5 g (Fig. 7.4). In the earlier field study of the genetic trade-off between current reproduction and future performance we found that current reproduction did not impact subsequent performance until capsule mass exceeded 0.6 g (Fig. 7.2). Given that capsules make up about 50% of reproductive biomass in *Plantago* (unpublished data), this is equivalent to a reproductive mass of 1.2 g; a value that is similar to that obtained using experimental manipulations to control level of reproduction.

The above two studies suggest the trade-off function between current and future reproduction in *Plantago* is nonlinear and convex. Further work is necessary before we can conclusively determine the shape of the trade-off function in these, as well as other species. However, there is sufficient evidence to justify an examination of some of the reasons why one might expect a convex trade-off function.

IV. Impact of Reproduction on Resource Uptake

As has been reviewed in Chapter 1 in this book, there is no clear separation between vegetative and reproductive functions in plants. Reproductive structures of many plants are active in photosynthesis and can supply a significant proportion of their own carbon requirements. Further, the induction of reproduction can stimulate rates of resource uptake in vegetative organs that can further reduce the extent to which reproductive structures are a drain on the rest of the plant. In other words, reproduction can increase the size of the total resource budget rather than simply taking a portion of a fixed resource budget (Bazzaz *et al.*, 2000). Given this increase in the resource budget no direct trade-off between reproductive and vegetative

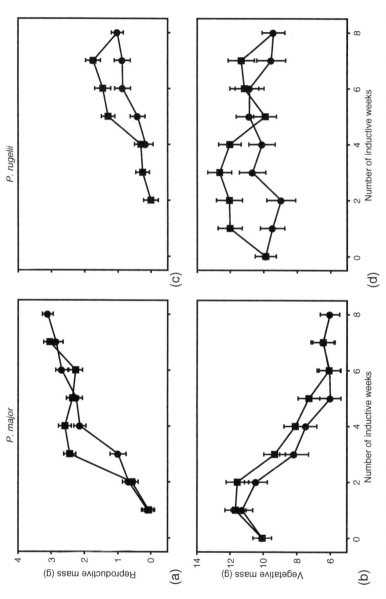

Figure 7.3 Reproductive and vegetative mass in *P. major* and *P. rugelii* as affected by the degree to which reproduction was induced (Reekie *et al.*, 2002). Induction was controlled by varying the number of weeks plants were exposed to an inductive photoperiod. Plants that were partially induced (1–7 weeks) were exposed to the inductive photoperiod either early (circles) or late (squares) in the experimental period. Error bars represent plus or minus one standard error.

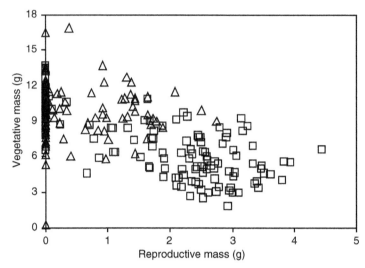

Figure 7.4 Vegetative mass plotted as a function of reproductive mass in *P. major* (squares) and *P. rugelii* (triangles) (Reekie *et al.*, 2002). Extent of reproduction was controlled experimentally by varying the number of weeks plants were exposed to an inductive photoperiod (see text). Each data point represents an individual plant. For a plot of vegetative and reproductive mass as function of the number of inductive weeks, see Fig. 7.3.

growth should be expected; i.e., the allocation of one unit of resources to reproduction should not result in a one unit decrease in vegetative growth. This by itself does not mean that the trade-off function between reproduction and growth is nonlinear; it could still be a linear relationship with a slope that is less than one providing the capacity of the plant to increase rates of resource uptake is a direct function of the resources invested in reproduction. However, if we assume that the capacity of the plant to increase rates of resource uptake with reproduction is limited, and that low levels of reproduction are sufficient to induce the maximum level of resource uptake, then the predicted trade-off would be a convex function (Fig. 7.1b). Why would plants have a limited capacity to increase rates of resource uptake in response to reproduction? We can offer two independent explanations for such a pattern.

The first explanation relates to the capacity of the reproductive structures to carry out photosynthesis and supply a portion of their own carbon needs. In most plants, the level of reproductive output is regulated at least in part by flower/fruit abortion (Willson, 1983; Lee, 1988; Lloyd, 1988; Stephenson, 1981). That is, plants generally produce more flowers than they do mature fruit; the excess flowers either never produce fruit, or the developing fruit are aborted at an early stage. This pattern is often interpreted as a form of bet-hedging in the face of an uncertain future.

The decision as to how many fruits are produced is delayed as long as possible in the event that the environment changes, either increasing or decreasing its favorability for reproduction. Given that the resources required for flower production are often insignificant relative to those required for seed/fruit maturation (Bazzaz *et al.*, 2000), the resource cost of delaying this decision may be relatively small. On the other hand, direct photosynthesis by the reproductive structures generally occurs at a higher rate in young developing inflorescences (Bazzaz and Carlson, 1979; Bazzaz *et al.*, 1979). As the reproductive structures mature, chlorophyll is degraded and many of the leaf-like structures responsible for this photosynthesis (e.g., sepals and bracts) senesce. Taken together, these observations suggest that as the level of reproductive output increases (i.e., a larger proportion of the flowers develop into fruits), the carbon cost will increase substantially, while the ability of the reproductive structures to compensate for this cost through photosynthesis will remain unchanged.

The second explanation relates to the stimulation of resource uptake in vegetative structures in response to reproduction. Such increases in resource uptake are usually interpreted as a consequence of increased sink strength (Bazzaz *et al.*, 2000). Rates of resource uptake under some circumstances have been shown to be limited by the capacity of the plant to utilize those resources rather than by the availability of these resources in the environment (Neales and Incoll, 1968; Wardlaw, 1990; Ho, 1992). Therefore, since reproduction increases resource demand or sink strength, it can increase resource uptake which may partially compensate for the resources invested in reproduction. However, it stands to reason that once resource demand increases beyond a certain point, the availability of resources in the environment will start to limit resource uptake (Bloom *et al.*, 1985) and there will be no further increases in resource uptake with increased reproduction. Further, the depletion of resources by reproduction may deprive leaves of necessary resources (e.g., nitrogen) lowering the photosynthetic capacity. Data on leaf photosynthesis from the experiment on *Plantago* described in the previous section, lend support to this hypothesis (Fig. 7.5). Low levels of reproductive allocation had highly variable effects on leaf photosynthesis, rates for reproductive plants often being higher or lower than those for vegetative controls. However, high levels of reproduction uniformly depressed photosynthesis.

V. Differences in the Resource Requirements of Vegetative versus Reproductive Tissue

Another reason why one might expect a convex trade-off function between reproduction and growth are differences in the resource requirements for the construction of vegetative versus reproductive tissue. Due to their

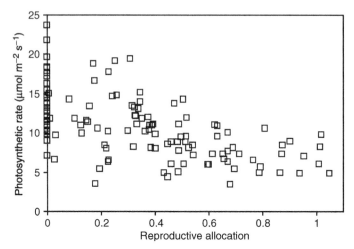

Figure 7.5   Photosynthetic capacity of the most recent fully emerged leaf in *P. major* plotted as a function of reproductive allocation (reproductive mass/vegetative mass) (Reekie *et al.,* 2002). Extent of reproduction was controlled experimentally by varying the number of weeks plants were exposed to an inductive photoperiod (see text).

different functions, vegetative and reproductive tissues often differ substantially in their resource requirements (Willson, 1983; Goldman and Willson, 1986). For example, in the case of *Plantago major,* reproductive tissue has a much higher requirement for nitrogen than vegetative tissue (Table 7.2). Similar differences are to be expected in their requirements for water (i.e., transpiration rates) and other mineral nutrients. Given that their resource requirements differ, this again means that one would not expect a direct trade-off between reproductive and vegetative growth. Further, the extent of the trade-off between reproduction and growth will vary depending upon which resources are most limiting. For instance, let us assume nitrogen availability is the most crucial factor limiting growth. In the case of *Plantago,* reproduction will significantly increase the demand for nitrogen per gram of tissue produced. Given that reproductive tissue requires almost three times as much nitrogen as vegetative tissue, the production of one gram of

Table 7.2   Nitrogen Content (Plus or Minus One Standard Deviation) of Vegetative (Leaf, Stem, and Root) and Reproductive Tissue (Spike with Capsules and Seeds) for *Plantago major* and *P. rugelii* (Unpublished Data)

Species	Vegetative tissue (% of dry weight)	Reproductive tissue (% of dry weight)
Plantago major	0.53 ± 0.08	1.61 ± 0.23
Plantago rugelii	0.55 ± 0.11	1.38 ± 0.60

reproductive tissue would require the sacrifice of almost three grams of vegetative tissue. Now, let us assume that nitrogen is abundant and some other resource for which reproduction and vegetative tissues have similar requirements is now limiting growth. One would now expect a one for one trade-off between reproduction and growth. Note that in both these examples, reproduction will increase the demand for nitrogen and nitrogen uptake will have to increase substantially to meet this demand. Therefore as the level of reproductive allocation increases, the rate of nitrogen uptake is more and more likely to limit growth and as it becomes more limiting, the trade-off between reproduction and growth will increase producing a convex trade-off function.

The above hypothesis seems logical, but unfortunately, is rather difficult to test empirically. An empirical test would require that we manipulate the resource requirements of vegetative versus reproductive tissue, as well as the availability of these resources in the environment. Manipulating resource availability in the environment is relatively straightforward. However, it is difficult to imagine how one might significantly alter the relative resource demands of vegetative versus reproductive tissue experimentally. Therefore, we constructed a growth and allocation model for *Plantago* to "test" this hypothesis (Reekie and Avila-Sakar, unpublished data). The model is based upon earlier models of allocation between roots and shoots described by Reynolds, Thornley and Grace (Reynolds and Thornley, 1982; Johnson and Thornley, 1987; Grace, 1997), but was modified to include reproductive allocation and to more realistically simulate resource uptake patterns.

The basic thrust of the model is that plant growth is limited by one of the two resources, carbon which is acquired by the leaves and nitrogen which is acquired by the roots (Fig. 7.6). Allocation of carbon and nitrogen uptake between leaf and root growth is determined by the carbon/nitrogen (C/N) ratio of the vegetative tissue. If the C/N ratio of the resources in the labile pool is less than that required for vegetative growth (i.e., there is a surplus of N), all available resources are allocated to leaf growth. On the other hand, if the C/N ratio in the labile pool is greater than that required for vegetative growth (i.e., there is a surplus of C), all available resources are allocated to root growth. The C/N ratio of the labile pool is determined by the relative uptake of C versus N, which in turn is determined by the availability of these resources in the environment, and the amount of root biomass for N and leaf area for C. The availability of carbon and nitrogen in the environment was considered constant for a given run of the model, but was varied among runs to provide four different environments (i.e., two levels of nitrogen availability crossed with two levels of carbon availability). Carbon availability was determined by irradiance; i.e., we assumed that the availability of light energy was the primary factor determining the rate of carbon uptake in photosynthesis. The uptake rate per unit root biomass or per unit leaf area was estimated by a rectangular hyperbola function which

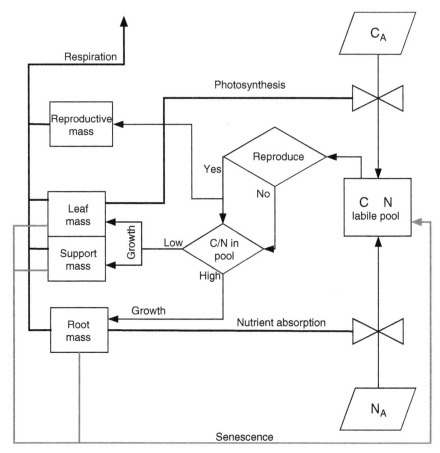

Figure 7.6   Flow diagram of the growth and allocation model used to examine the effect of variation in the C/N ratio of the reproductive tissue on the trade-off function between reproduction and vegetative growth in *Plantago*. See text for details.

adequately simulates the enzyme kinetics underlying the relationship between either nitrogen or irradiance and the rate of resource uptake (Grace, 1997). Light depletion within the canopy of individual plants (i.e., self-shading) was represented by an exponential decay function (i.e., the logarithm of irradiance declined in a linear fashion with leaf area index). Leaf area index in turn was a function of plant size and was estimated from allometric relationships derived from field collected plants of *Plantago major*. Leaf senescence was assumed to occur at the light compensation point (i.e., when photosynthesis = respiration). Respiration rate was considered a constant, so leaf senescence was effectively a function of light depletion within the canopy. Root senescence was a function of root age (life span = 30 days). We assumed 50% of the nitrogen in senescing plant parts was returned to the labile pool while none of the carbon was recycled. Allocation of resources to stem growth was based on the allometric relationship between stem biomass and the summed biomass of roots plus leaves for plants of *Plantago major.*

The above model was used to simulate the growth of vegetative plants over a 120-day growing season. To examine the effect of reproduction on growth, the model was modified to include allocation of an arbitrary proportion of the resources in the labile pool to reproductive growth after a vegetative growth period of 40 days. This proportion was varied among runs to simulate different levels of reproductive investment. The C/N ratio of the reproductive tissue was also varied among runs and was either identical to that of the vegetative tissue (80) or a value similar to that observed for reproductive tissue of *Plantago major* (30). On those runs in which reproductive tissue had a lower C/N ratio than vegetative tissue, allocation to reproduction changed the C/N ratio of the labile pool (i.e., it depleted the nitrogen content of the pool to a greater extent than vegetative growth alone).

To summarize the output of the model, the vegetative biomass attained at each level of reproductive allocation was plotted as a function of the resulting reproductive biomass for each of the four environmental conditions and for each of the three reproductive C/N ratios (Fig. 7.7). The average cost of reproduction for each level of reproductive investment was also calculated as the decrease in vegetative growth relative to nonreproductive plants divided by the reproductive biomass obtained (Table 7.3).

Depending upon the selected parameters, the model predicted either a rather steep trade-off function between vegetative and reproductive growth, or a trade-off function in which low levels of reproductive investment had little or no effect on vegetative growth, followed by a rapid decline in vegetative growth at higher levels of investment. However, the C/N ratio of the reproductive structures was not the primary factor responsible for variation in the shape of the function. This is evident when you look at the trade-off function for those runs in which the C/N ratio of the reproductive tissue was identical to that of the vegetative tissue (Fig. 7.7). If differences in the resource requirements for vegetative and reproductive growth were the primary explanation for the convex trade-off function, you would expect these relationships to be linear. However, even when the resource requirements of vegetative and reproductive growth were identical, it was still possible to have a curvilinear, convex relationship. This does not mean that the C/N ratio of the reproductive structures had no effect on reproductive cost, only that the difference in the C/N ratio between vegetative and reproductive tissue was not the primary factor responsible for the shape of the relationship. In fact, reducing the C/N of the reproductive structures from 80 (same as vegetative tissue) to 30 (normal ratio for reproductive tissue) almost doubled the cost of reproduction at the highest level

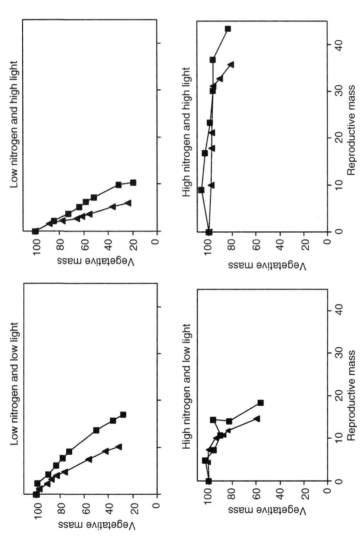

Figure 7.7   Vegetative mass plotted as a function of reproductive mass as predicted by the *Plantago* growth and allocation model (see text for details). The different levels of reproductive mass were obtained by varying the proportion (0.005, 0.01, 0.15, 0.02, 0.04, 0.06, 0.08, 0.1, or 0.2) of the labile resource pool invested in reproduction. Values are predicted for four different environments (2 levels of nitrogen availability crossed with 2 levels of light availability). Each graph displays two curves, one for reproductive tissue with the same C/N ratio (80) as vegetative tissue (squares), and one for reproductive tissue with a C/N ratio (30) typical of reproductive tissue (triangles).

Table 7.3 Average Slope of the Relationship between Vegetative and Reproductive Growth (See Fig. 7.7) as Affected by Resource Availability and C/N Ratio of the Reproductive Tissue. The Slopes Represent the Loss of Vegetative Growth That Results from One Gram of Reproductive Growth as Predicted by the *Plantago* Growth and Allocation Model (See Text for Details)

	C/N ratio of reproductive tissue	
Resource treatment	30	80
Low nitrogen and low light	−5.90	−3.65
Low nitrogen and high light	−12.06	−6.93
High nitrogen and low light	−2.74	−2.40
High nitrogen and high light	−0.51	−0.39

of reproductive investment in the low nitrogen, high light environment (Table 7.3). The C/N ratio of the reproductive structures also had a marked effect upon the reproductive cost in the low nitrogen, low light environment, but its effects in both high nitrogen environments were relatively minor. Given that, you would expect the nitrogen content of the reproductive structures to have more of an impact on reproductive cost when nitrogen is more limiting, this pattern is not unexpected. However, these changes in reproductive cost with reproductive C/N ratio were a result of changes in the slope or degree of curvature in the trade-off function rather than from a major transformation in the form of the relationship.

VI. Effect of Nitrogen versus Light Limitation

Given that the above model suggests the difference in the C/N ratio of reproductive versus vegetative tissue is not the primary factor responsible for the convex trade-off relationship in *Plantago*, we must look for alternative explanations. Examination of the predicted trade-off functions at different levels of light and nutrient availability may provide some insight into what this underlying cause might be. Although the trade-off function between vegetative and reproductive growth was nonlinear in all four environments, the degree of curvature varied dramatically among the environments. In the high nitrogen environments, the flat portion of the curve at low levels of reproductive investment was pronounced and was extended to relatively high levels of reproductive investment. At low nitrogen availability, the flat portion of the curve was limited to very low levels of reproductive invest-ment and was essentially absent in the case of the low nitrogen, high light environment (where one would expect nitrogen to be at its most limiting). We must therefore ask why nitrogen versus light limitation of growth results in such different trade-off functions.

The answer to this question appears to lie in the pattern of growth predicted by the model under nitrogen limitation. At high nitrogen availability, plant growth is rapid and the plant canopy quickly develops to the point where self-shading occurs. Self-shading provides a negative feedback upon growth and as a result, the growth curve, bears some superficial resemblance to a logistic growth curve, in that the rate of growth eventually starts to decline once the plant reaches a certain size (Fig. 7.8). Under severe nitrogen limitation however, growth rates are so slow that complete canopy closure does not occur within the length of the simulated growing season. As a result, self-shading does not provide negative feedback and the plants grew in a more or less exponential fashion over the study period (Fig. 7.8).

It is interesting to note that light limitation of growth did not result in the same growth pattern as nitrogen limitation, since it did not accentuate the exponential phase of the growth curve (Fig. 7.8). This occurred in spite of the fact that the low level of light chosen for this study was more limiting to final plant size than was the low level of nitrogen that was chosen (Fig. 7.8). Light limitation of growth is similar to that of nitrogen limitation as it also reduces plant size and therefore will tend to reduce the degree of self-shading (i.e., the number of layers of leaves through which the light must penetrate). However, it is quite different from nitrogen limitation, because decreases in light will also accentuate the impact of any shading that does occur. Given that decreases in light availability reduce the irradiance on the topmost leaf, the number of leaf layers required to reduce the irradiance to the light compensation point is reduced at low light availability. The net result of these two effects is that plants at low light exhibit a similar growth pattern (i.e., approximately logistic) as plants at high light.

VII. Effect of Growth Pattern

Upon inspection of the growth curves of the plants in the four environments, it is evident that differences in the trade-off functions among these environments (Fig. 7.7), are correlated with the differences in the growth curves (Fig. 7.8). In essence, the existence and length of the flat portion of the trade-off function in which low levels of reproductive investment have minimal impact on vegetative growth is associated with the slowing of growth due to self-shading. In retrospect, it is perhaps obvious that such a pattern should exist. After all, reproductive cost is calculated as the loss in vegetative growth when resources are allocated to reproduction. Therefore, if a plant invests only a small proportion of its resources to reproduction, the plant will grow relatively fast and reach a size at which self-shading starts to limit further growth. Modest increases in reproductive investment in this situation will have relatively little impact on vegetative growth as self-shading would have started limiting vegetative growth anyway once the plant reached

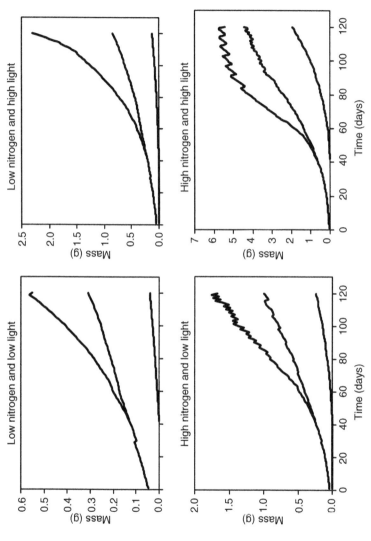

Figure 7.8 Total mass of a vegetative plant (top line), and vegetative (middle line) and reproductive mass (bottom line) of a reproductive plant as predicted by the *Plantago* growth and allocation model plotted as a function of time. The reproductive plant allocated 20% of its labile resource pool to reproduction. Values are predicted for four different environments (2 levels of nitrogen availability crossed with 2 levels of light availability). Note the change of scale on the vertical axis.

a certain size. However, at high levels of reproductive investment, or when self-shading is not an issue because growth rates are slow, investing resources in reproduction result in a sharp reduction in growth relative to plants that do not invest in reproduction and the calculated cost is relatively high. Thus, it seems that the main factor affecting the shape of the cost curve is the pattern of growth of the plant itself.

If the aforesaid assertion is correct, it should be possible to recreate the same trade-off functions with a much simpler model; one in which the only variables are level of reproductive investment and pattern of growth. We therefore used a simple discrete logistic equation to simulate growth:

$$M_{t+1} = \frac{M_t(1-I)R}{1+(M_t(1-I)(R-I)/K)}$$

Where M_t is the mass of the plant on day t, R is the growth rate (absolute rate of increase in biomass per unit biomass), K is the maximum size of the plant and I is the proportion of the labile pool devoted to reproduction. R, K, and I are constants. As in our first model, reproduction started after 40 days of vegetative growth and the length of the growing season was 120 days. This model does not explicitly include the effects of light versus nitrogen limitation, differences in C/N ratio, root versus shoot growth, self-shading, or rate of resource uptake as did the previous model, but it can produce growth curves that superficially resemble those produced by the previous model by choosing the appropriate values for growth rate (R) and maximum plant size (K). Using this model, we found that as the growth rate increases, the growth curve of the plant (biomass versus time) went from showing only the exponential phase to the whole sigmoid shape of a logistic curve (Fig. 7.9). This occurs because in the 120 days defined as the life span of a plant, plants with slower growth rates do not reach the maximum size proper of logistic growth, but they will reach it sooner with faster growth rates.

The most immediate effect of using part of the labile resource pool for the construction of reproductive tissues is to reduce the amount of resources available for vegetative growth, which in turn will slow down the rate of resource acquisition and the rate of vegetative growth. As a consequence, final vegetative biomass was a negative exponential function of I. The rate of this decrease in vegetative biomass with I was greater at lower growth rates. This result is a simple consequence of where the plant is on the growth curve. At low growth rates, the plant is still in the exponential growth phase, while at the highest growth rate, the plant is approaching the self-limiting portion of the growth curve and allocating resources to reproduction has less impact on vegetative growth.

The rate of accumulation of reproductive biomass depends on the biomass of vegetative tissues because of their capacity to garner resources from the

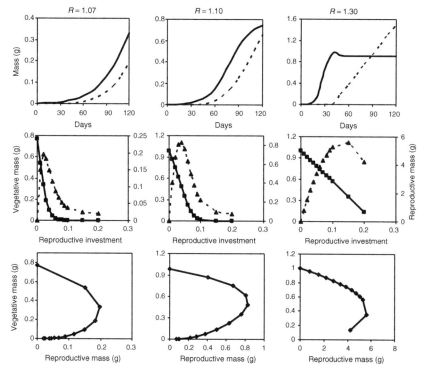

Figure 7.9 Predictions of a logistic growth model (see text for details) at three different growth rates *(R)*. The upper panel plots predicted vegetative (solid line) and reproductive mass (dashed line) as a function of time for a plant that allocates 5% of its daily growth to reproduction once it reaches day 40. The middle panel plots final (day 120) vegetative (squares) and reproductive mass (triangles) as functions of the proportion of daily growth allocated to reproduction. The lower panel plots final vegetative mass as a function of final reproductive mass for plants allocating increasingly larger proportions of their daily growth to reproduction. Note change in scale among graphs.

environment. Consequently, the increment of reproductive biomass per unit increment of reproductive investment becomes smaller as *I* increases, and will eventually become negative (Fig. 7.9). While the relation between reproductive biomass and reproductive investment always starts at zero and has a maximum, the magnitude of this maximum and the corresponding value of reproductive investment at which it is attained vary with the growth rate of the plant (Fig. 7.9). As the growth rate increases, the magnitude of the maximum reproductive biomass increases, and so does the corresponding value of reproductive investment at which the maximum is attained.

The cost function (i.e., the relationship between vegetative and reproductive biomass) had the same basic shape, regardless of growth rate (Fig. 7.9). Essentially, vegetative biomass declined as *I* and reproductive biomass increased from zero to the point of maximum reproductive biomass. Increasing I beyond this point resulted in further decreases in vegetative biomass as well as decreasing reproductive biomass. Eventually, increases in I drive both vegetative and reproductive biomass to zero. Given that natural selection is unlikely to favor increases in I beyond the point of maximum reproductive biomass, we can safely ignore that portion of the curve beyond this point. Hence, we end up with a convex trade-off function in which the cost of reproduction increases with level of reproductive investment. This function is very similar to that observed in the empirical studies of the cost of reproduction in *Plantago* described earlier. It would appear that a convex trade-off function is an inevitable consequence of the relationship between level of reproductive investment and the resultant reproductive mass. Given that unit increases in the level of reproductive investment result in successively smaller increments of reproductive mass due to the diversion of resources from vegetative structures and the impact this has upon future growth, a convex trade-off function is simply unavoidable.

Although growth rate did not change the basic shape of the trade-off function, it did have a marked impact upon the slope of the function. As the reproductive mass achieved with the same level of reproductive investment increases with growth rate, while the negative effect on vegetative growth declines, the average slope of the cost function decreases with increases in growth rate (Fig. 7.9). This means that the cost of reproduction decreases under conditions that favor rapid growth.

It should be noted that this latter prediction is dependent upon an important assumption that we make in this model. The model assumes that changes in growth rate have no impact on maximum plant size (K). If maximum plant size were to increase with growth rate, then the cost of reproduction could potentially remain constant or even increase. This is clear when you compare the results of this relatively simple model to those predicted by our earlier functional model. Increased nitrogen availability increased growth rate in the functional model (Fig. 7.8), and the resulting cost function (Fig. 7.7) decreased in slope regardless of whether light availability was low or high (Table 7.3). In this regard, the results of the functional model are very similar to those of the simplified logistic growth model (Fig. 7.9). However, increases in light availability increased growth rate in the functional model (Fig. 7.8), but did not necessarily decrease the cost of reproduction (Fig. 7.7). At high nitrogen availability, an increase in the light availability resulted in a decrease in the slope of the cost function, but at low nitrogen availability, an increase in light increased the slope of the cost function (Table 7.3). This latter observation is a direct result of the effect of light availability upon maximum plant size. Maximum plant size in the functional model is determined by the number of layers of leaves a plant can maintain before the lowest leaf reaches the light compensation point. Increased light availability means that more layers of leaves can be added before the lowest layer reaches the light compensation point (i.e., maximum plant size increases). At low nitrogen availability, growth is so limited by nitrogen that the plant never reaches maximum plant size and growth is exponential over the entire growth period (Fig. 7.8) resulting in a high cost of reproduction. On the other hand, at high nitrogen availability, the plant grows fast enough to approach the maximum plant size by the end of the growth period (Fig. 7.8), and the cost of reproduction declines with increased light availability.

VIII. Conclusions

The difficulties associated with accurately assessing the cost of reproduction have largely precluded an accurate assessment of the shape of the trade-off function between reproduction and growth in plants. It is important that we know what this trade-off function is like, if we are to understand how selection will act upon variation in reproductive allocation. In this chapter we have presented empirical evidence that in *Plantago,* this function is curvilinear and convex. A convex function is an inevitable consequence of the relationship between reproductive allocation and reproductive output. Increasing resource allocation to reproduction initially increases reproductive output. Eventually, however, increases in reproductive allocation will lead to decreased reproductive output due to the negative impact of this allocation on vegetative growth and consequently, the size of the available resource pool (i.e., reproductive output is the product of reproductive allocation and the size of the resource pool). As a result, when vegetative growth is plotted against reproductive output, the relationship will always curve sharply downwards if high enough levels of reproductive output are achieved. That said, there are a number of other factors that can modify this relationship, increasing or decreasing the slope or degree of curvature in this relationship. This may result in trade-off functions with a slope of zero (i.e., no reproductive cost) or functions which appear linear, if there is insufficient variation in reproductive output and only a portion of the relationship is plotted. Such factors include the extent to which reproduction can enhance photosynthesis, differences in the resource requirements of vegetative versus reproductive tissue, and the growth pattern of the plants involved.

The capacity of a plant to increase photosynthesis in response to reproduction either through direct photosynthesis by reproductive structures, or by enhancing leaf photosynthesis, is likely to be limited for a variety of reasons. Therefore, although the carbon cost of limited reproduction may be largely compensated for by this enhanced carbon gain, higher levels of reproductive output will likely entail much greater costs. This effect will accentuate the pattern described above, enhancing the convex nature of the trade-off function.

The effect of differences in the resource requirements of vegetative versus reproductive growth on the shape of the trade-off function varies depending upon resource availability. If the resource which is required in greater amounts by reproductive structures has limited availability, the slope of the trade-off function and degree of curvature will increase. On the other hand, it will have minimal effect on the trade-off function at higher levels of resource availability.

Growth pattern can have a marked influence on the shape of the trade-off function. Increases in growth rate and decreases in maximum plant size will tend to reduce the cost of reproduction and will result in a trade-off function with a very shallow slope until very high levels of reproductive output are achieved, at which point there is a sharp increase in cost. In essence, vegetative growth may be constrained or limited by the architecture or morphology of the plant, given the set of environmental conditions in which the plant is found. If reproductive growth is free of these constraints, for example, reproductive structures may have a minimal impact on self-shading due to a different morphology as in *Plantago,* allocation of resources to reproduction will have minimal or no impact upon vegetative growth. It follows then that to better understand the factors determining reproductive cost, we have to better understand the factors limiting vegetative growth in particular environments, and to what extent reproductive growth may or may not avoid these constraints.

Acknowledgments

The research reported in this study was supported by the Natural Science and Research Council of Canada.

References

Ågren J. and Willson M. F. (1994) Cost of seed production in the perennial herbs *Geranium maculatum* and *G. sylvaticum:* An experimental field study. *Oikos* **70**: 35–42.

Bailey R. C. (1992) Why we should stop trying to measure the cost of reproduction correctly. *Oikos* **65**: 349–352.

Bazzaz F. A. and Carlson R.W. (1979) Photosynthetic contribution of flowers and seeds to reproductive effort of an annual colonizer. *New Phytologist* **82**: 223–232.

Bazzaz F. A., Carlson R. W. and Harper J. L. (1979) Contribution to reproductive effort by photosynthesis of flowers and fruits. *Nature* **279**: 554–555.

Bazzaz F. A., Ackerly D. D. and Reekie E. G. (2000) Reproductive allocation in plants. In *Seeds: The Ecology of Regeneration in Plant Communities* (M. Fenner, ed.) pp. 1–29, CABI Publishing, Wallingford.

Bloom A. J., Chapin F. S. III and Mooney H. A. (1985) Resource limitation in plants - an economic analogy. *Annual Review of Ecology and Systematics* **16**: 363–392.

Calvo R. N. and Horvitz C. C. (1990) Pollinator limitation, cost of reproduction, and fitness in plants: A transition matrix demographic approach. *American Naturalist* **136**: 499–516.

Cipollini M. L. and Whigham D. F. (1994) Sexual dimorphism and cost of reproduction in the dioecious shrub *Lindera benzoin* (Lauraceae). *American Journal of Botany* **81**: 65–75.

Farris M. A. and Lechowicz M. J. (1990) Functional interactions among traits that determine reproductive success in a native annual plant. *Ecology* **71**: 548–557.

Fox J. F. and Stevens G. C. (1991) Costs of reproduction in a willow: Experimental responses versus natural variation. *Ecology* **72**: 1013–1023.

Geber M. A. (1990) The cost of meristem limitation in *Polygonum arenastrum*: Negative genetic correlations between fecundity and growth. *Evolution* **44**: 799–819.

Goldman D. A. and Willson M. F. (1986) Sex allocation in functionally hermaphroditic plants: a review and critique. *Botanical Review* **52**: 157–194.

Grace J. (1997) Toward models of resource allocation by plants. In *Plant Resource Allocation* (F.A. Bazzaz, and J. Grace, eds.) pp. 279–291. Academic Press, San Diego.

Hawthorn W. R. (1974) The biology of Canadian weeds. 4. *Plantago major* and *P. rugelli*. *Canadian Journal of Plant Science* **54**: 383–396.

Hawthorn W. R. and Cavers P. B. (1976) Population dynamics of the perennial herbs *Plantago major* L. and *P. rugelli* Decne. *Journal of Ecology* **64**: 511–527.

Horvitz C. C. and Schemske D. W. (1988) Demographic cost of reproduction in a neotropical herb, an experimental approach. *Ecology* **69**: 1741–1745.

Ho L. C. (1992) Fruit growth and sink strength. In *Fruit and Seed Production: Aspects of Development, Environmental Physiology and Ecology* (C. Marshall and J. Grace, eds.) pp. 101–124. Cambridge University Press, Cambridge.

Jennersten O. (1991) Cost of reproduction in *Viscaria vulgaris* (Caryophyllaceae): A field experiment. *Oikos* **61**: 197–204.

Johnson I. R. and Thornley J. H. M. (1987) A model of shoot: root partitioning with optimal growth. *Annals of Botany* **60**: 133–142.

Karlsson P. S., Svensson B. M., Carlsson B. A. and Nordell K. O. (1990) Resource investment in reproduction and its consequences in three *Pinguicula* species. *Oikos* **59**: 393–398.

Kozlowski T. T. (1971) *Growth and Development of Trees, Vol. 2*. Academic Press, New York.

Law R. (1979) The cost of reproduction in annual meadow grass. *American Naturalist* **113**: 3–16.

Lee T. D. (1988) Patterns of fruit and seed production. In *Plant Reproductive Ecology – Patterns and Strategies* (J. Lovett Doust and L. Lovett Doust, eds.) pp. 179–202. Oxford University Press, New York.

Lloyd D. G. (1988) Benefits and costs of biparental and uniparental reproduction in plants. In *The Evolution of Sex* (R. E. Michod and B. R. Levin, eds.) pp. 233–252. Sinauer Assoc., Sunderland, Massachusetts.

Mitchell-Olds T. and Bergelson J. (1990) Statistical genetics of an annual plant, *Impatiens capensis*. II. Natural selection. *Genetics* **124**: 417–421.

Neales T. F. and Incoll L. O. (1968) The control of leaf photosynthesis rate by the level of assimilate concentration in the leaf: A review of the hypothesis. *Botanical Review* **34**: 107–125.

Newell E. A. (1991) The direct and delayed costs of reproduction in *Aesculus californica*, the California buckeye tree. *Journal of Ecology* **79**: 365–378.

Pfister C. A. (1992) Costs of reproduction in an intertidal kelp: patterns of allocation and life history consequences. *Ecology* **73**: 1586–1596.

Piñero D., Sarukhán J. and Alberdi P. (1982) The costs of reproduction in a tropical palm, *Astrocaryum mexicanum*. *Journal of Ecology* **70**: 473–481.

Primack R. B. and Hall P. (1990) Costs of reproduction in pink lady's slipper orchid: a four year experimental study. *American Naturalist* **136**: 638–656.

Primack R. B., Miao S. L. and Becker K. R. (1994) Costs of reproduction in the pink lady's slipper orchid (*Cipripedium acaule*): defoliation, increased fruit production and fire. *American Journal of Botany* **81**: 1083–1090.

Ramadan A. A., El-Keblawy A., Shaltout K. H. and Lovett-Doust J. (1994) Sexual polymorphism, growth, and reproductive effort in Egyptian *Thymelaea hirsuta* (Thymelaeceae). *American Journal of Botany* **81**: 847–857.

Ramsey M. (1997) No evidence for demographic costs of seed production in the pollen-limited perennial herb *Blandfordia grandiflora* (Liliaceae). *International Journal of Plant Sciences* **158**: 785–793.

Reekie E. G. (1998a) An explanation for size-dependent reproductive allocation in *Plantago major*. *Canadian Journal of Botany* **76**: 43–50.

Reekie E. G. (1998b) An experimental field study of the cost of reproduction in *Plantago major* L. *Ecoscience* **5**: 200–206.

Reekie E. G. and Bazzaz F. A. (1992) Cost of reproduction in genotypes of two congeneric plant species with contrasting life histories. *Oecologia* **90**: 21–26.

Reekie E. G., Budge S. and Baltzer J. L. (2002) The shape of the trade-off function between reproduction and future performance in *Plantago major* and *Plantago rugelii*. *Canadian Journal of Botany* **80**: 140–150.

Reekie E. G. and Reekie J. Y. C. (1991) An experimental investigation of the effect of reproduction on canopy structure, allocation and growth in *Oenothera biennis*. *Journal of Ecology* **79**: 1061–1071.

Reynolds J. F. and Thornley J. H. M. (1982) A shoot:root partitioning model. *Annals of Botany* **49**: 585–597.

Reznick D. (1985) Costs of reproduction, an evaluation of the empirical evidence. *Oikos* **44**: 257–267.

Saikkonen K., Koivunen S., Vuorisalo T. and Mutikainen P. (1998) Interactive effects of pollination and heavy metals on resource allocation in *Potentilla anserina* L. *Ecology* **79**: 1620–1629.

Schmid B. (1990) Some ecological and evolutionary consequences of modular organization and clonal growth in plants. *Evolutionary Trends in Plants* **4**: 25–34.

Schwaegerle K. E. and Levin D. A. (1990) Environmental effects on growth and fruit production in *Phlox drummondii*. *Journal of Ecology* **78**: 15–26.

Stearns S. C. (1989) Trade-offs in life-history evolution. *Functional Ecology* **3**: 259–268.

Stephenson A. G. (1981) Flower and fruit abortion: proximate causes and ultimate functions. *Annual Review of Ecology and Systematics* **12**: 253–279.

Syrjanen K. and Lehtilä K. (1993) The cost of reproduction in *Primula veris*: differences between two adjacent populations. *Oikos* **67**: 465–472.

Thoren L. M., Karlsson P. S. and Tuomi J. (1996) Somatic cost of reproduction in three carnivorous *Pinguicula* species. *Oikos* **76**: 427–434.

Wardlaw I. F. (1990) The control of carbon partitioning in plants. *The New Phytologist* **116**: 341–381.

Williams G. C. (1966) Natural selection, the cost of reproduction and a refinement of Lack's principle. *Am. Nat.* **100**: 687–690.

Willson M. F. (1983) *Plant Reproductive Ecology*. Wiley, New York.

Zimmerman J. K. (1991) Ecological correlates of labile sex expression in the orchid *Catasetum viridiflavum*. *Ecology* **72**: 597–608.

8

On Size, Fecundity, and Fitness in Competing Plants

Lonnie W. Aarssen

I. Introduction

An amateur naturalist knows that competitive exclusion during succession in vegetation usually involves success for larger species at the expense of smaller ones. Accordingly, traditional competition theory assumes that stronger competitive ability in plants requires larger size (e.g., Grime, 1979; Keddy, 1989) and that differences in size, combined with intense competition generally leads to competitive exclusion (Fig. 8.1). The same amateur naturalist also knows, however, that coexisting species usually have a wide range of sizes at virtually all stages of succession. Size distributions are, in fact, usually right skewed, with most species being relatively small (Aarssen and Schamp, 2002). Although these observations appear conflicting at first, they are reconciled by another assumption of traditional competition theory: the coexistence of species, including species of different sizes, implies that they compete only weakly, or not at all (Tokeshi, 1999). Traditional theory further identifies two main mechanisms that can promote weak competition between species and hence, coexistence: (i) *frequent or regular disturbance* that maintains community density below equilibrium, and/or community biomass below carrying capacity (or limits the time over which equilibrium/carrying capacity is attained) (Connell, 1978; Grime, 1979; Huston, 1979), which may be combined with repeated colonization of disturbance-generated gaps by immigrants from neighboring communities; and (ii) *narrow niche overlap,* allowing species to avoid some measure of

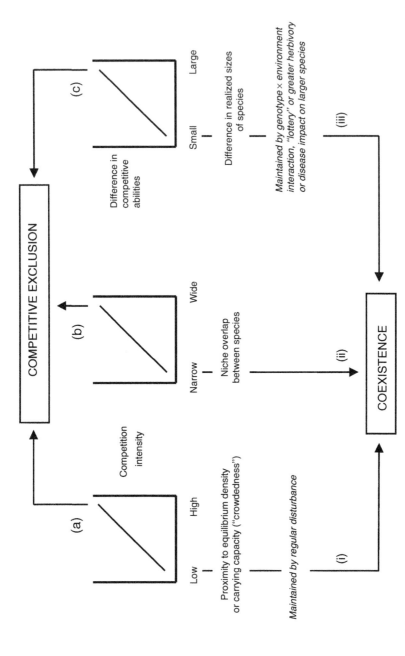

Figure 8.1 Traditional theory defines three requirements for competitive exclusion – high "crowdedness" (a), and wide niche overlap (b) (both promoting high competition intensity), plus a large difference in competitive abilities (c) – and three corresponding corollaries representing three general mechanisms promoting species coexistence in vegetation (see text).

interaction by having fundamental requirements for, and hence by making demands on, different resource units (e.g., differences between species in seed germination requirements, seedling establishment requirements, depth and placement of roots, soil moisture/nutrient requirements, phenology, use of pollinators) (Grubb, 1977; Harper, 1977; Tilman, 1982; Tokeshi, 1999) (Fig. 8.1).

The more astute amateur naturalist will also observe that some relatively small species can be found coexisting with species that have potential for larger size but which, possibly because of this potential to be particularly conspicuous, are sometimes disproportionately attacked by consumers and hence, suppressed in size. Species may coexist, therefore, despite intense competition (i.e., even at carrying capacity and without niche differentiation), provided that differences in sizes (competitive abilities) are, on average, not realized (at least not for long) because of differential consumer effects (e.g., Harper, 1969; Janzen, 1970; Fox, 1977; Risch and Carroll, 1986). Modern theory includes this effect under a third general mechanism for species coexistence: (iii) coexistence is possible even under intense competition, but only if there are *small differences in competitive abilities* (Fig. 8.1). Most of the theory involving this mechanism assumes that realized sizes (i.e., competitive abilities) are affected by the abiotic environment and that these environmental effects are heterogeneous in space and/or time. Hence, several coexisting species may all have the same fundamental niche requirements defining the same array of sites within a habitat where they could each leave descendants in the *absence* of competing species. However, in the *presence* of competitors at equilibrium density or carrying capacity, each species may "take turns" being the superior competitor (i.e., with the larger or largest size) in different microsites or at different times (e.g., seasons) within the habitat. The "winner" of local competition then is determined by "lottery" when the environmental heterogeneity that affects competitive ability (size) occurs at the small scale of a single individual (e.g., Agren and Fagerstrom, 1984; Fagerstrom, 1988). Alternatively, the "winner" is determined by species (or genotype) × environment interaction when the environmental heterogeneity that affects competitive ability occurs at the larger scale of the local neighborhood (e.g., Steward and Levin, 1973; Tilman, 1984, 1986; Chesson, 1986; Ellner, 1987; Aarssen, 1992; Bengtsson *et al.*, 1994).

The most discerning of amateur naturalists will notice further, however, that some large plant species (e.g., *Acer saccharum* in eastern deciduous forests of North America), during succession, can competitively exclude other species that are equally large or even larger (e.g., *Populus balsamifera* and *Populus deltoides*) (Burns and Honkala, 1990). Some species (e.g., *Faxinus nigra*) can be competitively excluded by species that are much smaller (e.g., *Ostrya virginiana* and *Carpinus caroliniana*) (Delcourt and Delcourt, 1987). In some cases, therefore, stronger competitive ability in plants clearly involves traits other than, or in addition to larger size. Yet, recent interpretations of competitive ability in plants remain firmly entrenched in the assumption that differences in competitive abilities are determined primarily by differences in plant size (e.g., biomass, height, leaf area) or in the traits that promote differences in plant size (e.g., Gaudet and Keddy, 1988; Grace, 1990; Keddy, 1990; Goldberg, 1996; Freckleton and Watkinson, 2000; Connolly *et al.*, 2001).

This bias in favor of plant size comparisons is strongly reflected in empirical data on competitive ability measurement from the published literature (Aarssen and Keogh, 2002), most of which regards competitive ability in terms of the traits that cause the competitive exclusion of species that lack them. Indeed, differences in plant size probably account for most cases of competitive exclusion in natural vegetation. There is, however, a different sense in which competitive ability plays an equally important but largely neglected role: in terms of the traits that, under intense competition, *prevent* the competitive exclusion of species that *possess* them. If differences in competitive abilities, causing competitive exclusion, are usually accounted for by differences in plant size, does it follow then, that *similarity* in competitive abilities, allowing *coexistence,* is usually accounted for by *similarity* in plant size (Fig. 8.1)? Below, it is argued that this is not likely to be generally true because more than just size matters. Some species may indeed coexist because of similarity in size. However, if more than size matters in affecting competitive ability, then this raises the interesting possibility that species may coexist not only because they might compete weakly or because they might "take turns" being the superior (larger) competitor in a heterogeneous environment, but also (or rather) because they might compete intensely, but with similar abilities regardless of how heterogeneous the environment is and despite differences in sizes. This would have significant implications for both theory development in the interpretation of diversity patterns across environmental gradients, as well as for applications involving the preservation and restoration of biodiversity in natural habitats.

Underlying assumptions for this mechanism of species coexistence are explored in this chapter. The focus is on several key questions: How many component traits of competitive ability are there for plants? How do these components interact with each other in preventing the competitive exclusion of plants that possess these traits? When are some components just as, or more important than plant size? These questions are addressed subsequently by developing a simple conceptual model, in which several component traits of competitive ability (including size) interact as allocation trade-offs, and in combinations that define several alternative ways of having small differences in competitive abilities between-species and hence, permit coexistence of different-sized plants across generations, despite intense competition between them (i.e., despite "crowding" with widely overlapping niche requirements), and without requiring differential consumer effects or a heterogeneous physical environment.

II. Defining the Components of Competitive Ability for Between-species Plant Competition: Lessons from Within-species Competition

For a single population at equilibrium density, where births are balanced by deaths and biomass is at carrying capacity (i.e., competition is intense), we know, by definition, that the average plant leaves exactly one descendant. Yet, as Harper (1977) suggested, "it is an intriguing paradox that natural selection continually favors individuals that leave more." We know, therefore, that high fitness (number of descendants) within this population is determined ultimately by relatively high lifetime output of reproductive offspring (i.e., offspring that become reproductive), despite intense lifetime intraspecific competition. Other traits (e.g., large size) may contribute but ultimately, "natural selection recognizes only one currency... offspring production" (Pianka, 1988). The benefit of producing more than one reproducing offspring (higher fitness) is always enjoyed exclusively by the parent plant. In natural vegetation, however, as in Hardin's (1968) "tragedy of the commons," the cost of this extra offspring production – increased crowding – is always shared with other neighboring plants. Hence, the per capita magnitude [benefit – cost] is always greater for individuals that produce more than one reproductive offspring than for neighboring individuals in the same population that produce only one. Traditional Malthusian theory predicts, therefore, that natural selection will always favor those individual traits that promote high lifetime output of reproductive offspring relative to neighbors, even among severely impoverished individuals that are already intensely crowded at or above carrying capacity (Hardin, 1968, 1993; Endler, 1986).

Extending these well-established principles of population biology to multispecies vegetation gives two inevitable predictions for plant competition and species coexistence theory: (i) several species can coexist at equilibrium (with community biomass at carrying capacity) and with similar niches (i.e., despite intense competition), provided that the average plant of *each* species leaves no more *and no less* than one descendant; and (ii) fitness (number of descendants) within this multispecies community of competitors will be determined directly by the relative ability to accrue reproductive offspring over the generation time of the longest-lived competitor, despite intense competition over this time, both within- *and* between-species.

The components of relative competitive ability within multispecies vegetation at equilibrium density or carrying capacity, therefore, can and should be interpretable in terms of the same components that define relative competitive ability within a single species population at equilibrium density/carrying capacity. Accordingly, strong competitive ability can be meaningfully defined only as high fitness under competition, i.e., high reproductive offspring output across generations, despite intense competition. In terms of measurable traits, this is estimated by relatively high fecundity under competition over at least one complete generation. The critical question then becomes: what are the principal fitness components contributing to high lifetime fecundity under competition? Following traditional theory of Darwinian fitness, there should be three: high *growth* under competition (i.e., ability to deny resources to competitors), high *survival* (longevity) under competition (i.e., ability to tolerate resource denial by competitors), and high *reproduction* under competition (i.e., high fecundity per unit plant size per unit time, when plant size and/or time for reproduction are limited by competition) (Fig. 8.2) (Aarssen, 1983, 1989, 1992; Aarssen and Keogh, 2002). Plant size, therefore, although clearly important, can only be regarded as one of the three fundamental components of competitive ability in plants, one of which, fecundity per unit plant size per unit time (Aarssen and Taylor, 1992), has been virtually ignored in empirical studies (Aarssen and Keogh, 2002).

III. Predicting Fecundity under Competition

Variation in the fecundity of seed plants under competition can be predicted from a simple allocation model (Fig. 8.3). The model developed, below, predicts variation in female fecundity (i.e., number of seed offspring matured per plant) and assumes that pollen : ovule ratio is constant, and that female fecundity is not pollen- or pollinator-limited. Predictions for variation in male fecundity (number of seeds sired per plant) or variation in allocation to male versus female fecundity is left for consideration elsewhere.

There are four stages in the model affected by three distinct sources of phenotypic variation: age, environment, and genotype (or species identity):

(i) *Variation in number of seeds per plant (fecundity, F)* is most proximately a function of variation in [total seed mass]/[single seed mass]:

$$F = \left(\frac{\text{total seed mass}}{\text{single seed mass}} \right)$$

Variation in single seed mass is determined primarily by genetic variation (Fig. 8.3), but may also be affected to some extent by environmental or maternal effects (Roach and Wulff, 1987).

(ii) *Variation in total seed mass per plant* is most proximately a function of variation in total plant mass. The effect of total plant mass is modulated primarily by genetic variation affecting both variation in

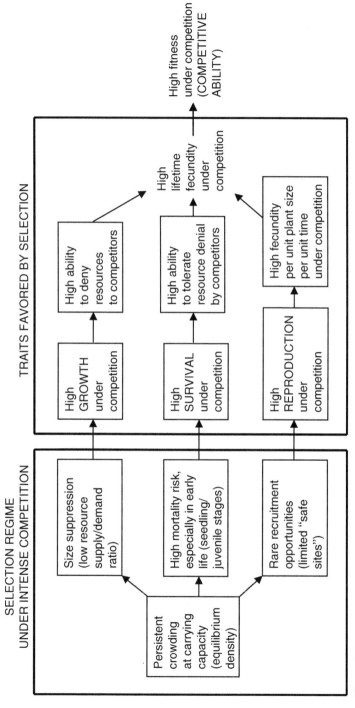

Figure 8.2   The selection regime under intense competition within vegetation defines three principal traits that may be favored by natural selection and hence, three corresponding fundamental components of competitive ability in plants (see text).

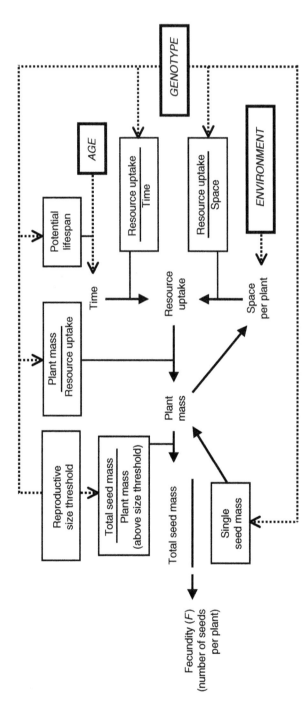

Figure 8.3 Path diagram for predicting variation in female fecundity in seed plants (number of seed offspring matured per plant) under competition. Dotted arrows leading to individual traits indicate predicted effects from one of three distinct sources of phenotypic variation: age, environment, and genotype (or species) (see text).

reproductive size threshold (the minimum plant mass required for successful seed production) (Fig. 8.3) and variation in total seed mass per unit plant mass (Fig. 8.3) (e.g., Reekie and Bazzaz, 1987):

$$F = \left(\frac{\text{total seed mass}}{\text{single seed mass}} / \text{plant mass} \right) \text{plant mass}$$

(iii) *Variation in total plant mass* is most proximately a function of variation in resource uptake, but may also be affected by "initial" plant mass, as determined by individual seed mass (Fig. 8.3). The effect of resource uptake is modulated by genetic variation affecting variation in plant mass accumulation per unit resource uptake (Fig. 8.3) (e.g., Berendse and Aerts, 1987; Nijs and Impens, 2000):

$$F = \left(\frac{\text{total seed mass}}{\text{single seed mass}} / \text{plant mass} \right)$$
$$\times [\text{plant mass}/\text{resource uptake}] \text{resource uptake}$$

(iv) *Variation in resource uptake* is a direct function of variation in time elapsed and variation in available space per plant (for a given background level of resource availability, e.g., soil fertility). The effect of time is modulated by genetic variation affecting variation in resource (e.g., nutrient) uptake per unit time (e.g., Chapin, 1980). Similarly, the effect of space per plant will be modulated by genetic variation affecting variation in resource (e.g., nutrient) uptake per unit space (e.g., per unit volume of soil) (Fig. 8.3) (e.g., as reflected in the "R*" of Tilman (1988)):

$$F = \left(\frac{\text{total seed mass}}{\text{single seed mass}} / \text{plant mass} \right)$$
$$\times \left(\text{plant mass}/\frac{\text{resource uptake}}{\text{time}} \times \text{time} \right)$$
$$\times \left(\frac{\text{resource uptake}}{\text{space}} \times \frac{\text{space}}{\text{plant}} \right)$$

Variation in time elapsed represents variation in age. Variation in the effect of age will be modulated by genetic variation affecting variation in potential life span (Fig. 8.3). Realized life span will also be affected by environmental variation (not shown in Fig. 8.3) and variation in age will also be affected by variation in germination time (not shown in Fig. 8.3), which in turn may be a function of genetic or environmental variation (Baskin and Baskin, 1998), or chance (Bosey and Aarssen, 1995). *Variation in space per plant* is effected by variation in plant density (local proximity of neighbors) which represents the principal environmental variable in the present model (Fig. 8.3); (the inclusion of other important environmental variables, such as variation in soil fertility, could also be considered but is not pursued here). Space available per plant is also affected, however, by plant size (Fig. 8.3) and in particular, plant size *relative* to neighbors; plants that are larger than their neighbors occupy more space and hence, as they grow, deny more space, usually *disproportionately* more space to these neighbors (Harper, 1977).

IV. Relationships among Plant Traits Affecting Fecundity under Competition: Alternative Ways to Compete Intensely While Avoiding Competitive Exclusion

Based on the allocation model as described earlier (Fig. 8.3), "cascading" graph diagrams can be used to model several series of cause–effect relationships among plant traits that predict variation in fecundity between individuals. The effects of variation in age alone (Fig. 8.4a) or environment (space per plant) alone (Fig. 8.4b) are straightforward: older and/or less crowded plants are larger and hence, more fecund. When they both vary, however, the effects of age and environment may interact. In a study using *Arabidopsis thaliana* (Clauss and Aarssen, 1994a), the increase in plant mass and fecundity with age was positively allometric (scaling coefficient, $b > 1.0$) in more favorable environments ($1'-5'$ in Fig. 8.5a) but isometric ($b = 1.0$) in less favorable environments ($1-5$ in Fig. 8.5a). Similarly, the increase in plant mass and fecundity resulting from environmental amelioration was negatively allometric ($b < 1.0$) at young ages ($1-1'$ in Fig. 8.5a) but isometric at maturity (final developmental stage) ($5-5'$ in Fig. 8.5a). The relationship between plant size and fecundity may also depend on which environmental variable (e.g., variation in light versus water versus soil nutrients) is the primary cause of variation in plant size (Clauss and Aarssen, 1994b) (Fig. 8.5b).

Predictions for fecundity variation are also straightforward based on effects of genetic variation alone when it is restricted only to those traits that affect variation in plant size; i.e., all else being equal, plants with greater resource uptake rate (per unit space or per unit time) or greater plant growth rate (mass per unit resource uptake) are also larger and hence, more fecund (Fig. 8.6). Predictions are again straightforward if

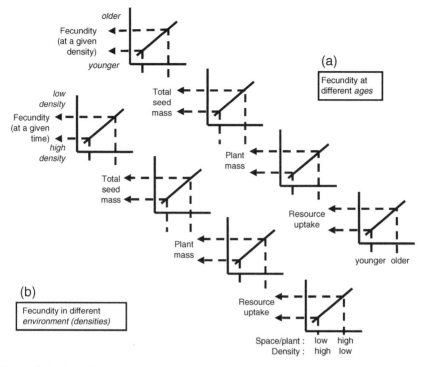

Figure 8.4 Cascading graph diagram for predicting variation in fecundity as a consequence of: (a) variation in age; or (b) variation in environment (space per plant).

genetic variation is restricted only to those traits that affect variation in survival or longevity (Fig. 8.7a), or only those traits that affect variation in reproductive size threshold (Fig. 8.7b), variation in total seed mass per unit vegetative mass (Fig. 8.7c), or variation in fecundity per unit seed mass (Fig. 8.7d). If total plant mass (P) is constant, then total seed mass (R) and vegetative mass (V) trade-off directly across genotypes (Fig. 8.7c). Similarly, if total seed mass (R) is constant, then fecundity (F) and individual seed mass (S) trade-off directly across genotypes (Fig. 8.7d).

Note that if genotypes or species have different longevities, then the relevant time scale for fitness comparison must encompass the greater (or greatest) lifespan. The important fecundity comparison, therefore, will include total offspring production (e.g., germinable seed) across the time interval encompassing all generations of the shorter-lived plant that are completed within the lifespan of the longer-lived plant. Unless viability of dormant seed is of sufficient duration for the shorter-lived plant, however, many or most of these offspring will perish because of the longer duration of space occupancy by the longer-lived species.

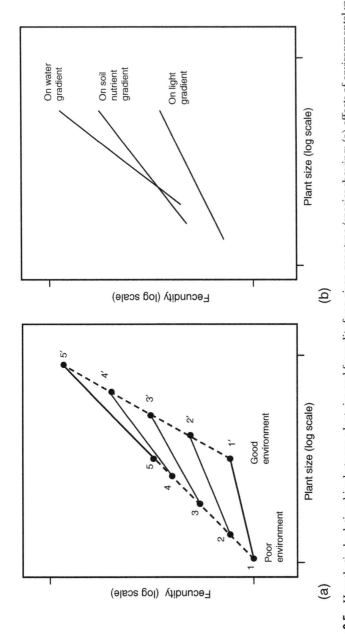

Figure 8.5 Hypothetical relationships between plant size and fecundity for a given genotype/species showing: (a) effects of environmental variation (solid lines) at different ages (1–5) with effect of age variation shown as dashed lines in poor versus good environments (see text); and (b) that size–fecundity relationships may vary (i.e., with different slopes) depending on the type of environmental variable (e.g., light, water, soil nutrients) that is causing variation in plant size (at a given age) (see text).

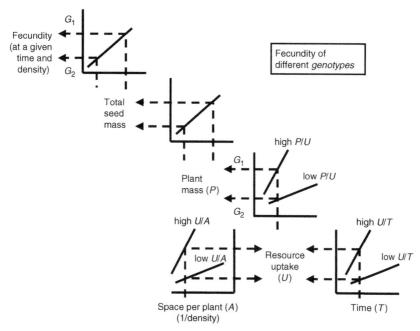

Figure 8.6   Cascading graph diagram for predicting variation in fecundity as a consequence of genotypic variation (G_1 versus G_2) in traits that affect variation in plant size, i.e., resource uptake per unit space (U/A), resource uptake per unit time (U/T), or plant growth (mass) per unit resource uptake (P/U).

Note also that for each of the pairwise plant comparisons earlier, differences in fecundity occur regardless of whether the two plants (e.g., younger versus older, slower versus faster resource uptake, shorter- versus longer-lived, lower versus higher reproductive allocation) are actually competing with each other. If they are competing, however, any associated differences in plant size are expected, through asymmetric dominance/suppression effects (Harper, 1977), to lead to greater differences in space per plant (Fig. 8.3) leading to even greater differences in size and in turn, greater differences in fecundity and hence, greater differences in competitive ability in Figs 8.4–8.7.

Normally, we would expect the effects of genetic variation to be more complicated than in the aforesaid considerations, not only because these effects will depend on which traits are genetically variable (which may differ between species), but also because the effects of genetic variation will depend on the way that genetically variable traits are correlated, both within-and between-species. Accordingly, cascading graph diagrams can be used to illustrate the prediction that "size" alone will often not be the only thing that matters. If plant size or more specifically, plant mass per unit

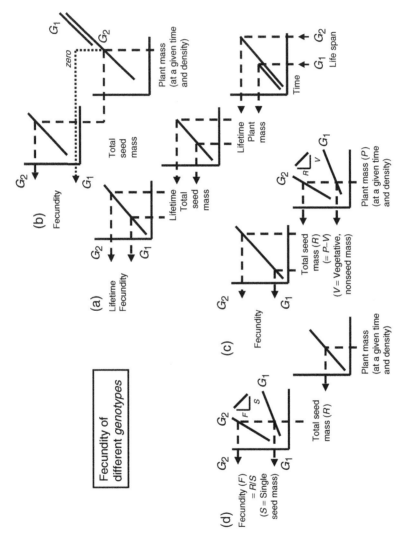

Figure 8.7 Cascading graph diagram for predicting variation in fecundity as a consequence of genotypic variation (G_1 versus G_2) in: (a) survival or longevity; (b) reproductive size threshold; (c) total seed mass (R) per unit vegetative mass (V) where R and V trade-off directly across genotypes; or (d) fecundity (F) per unit seed mass (S) where F and S trade-off directly across genotypes.

resource uptake (or resource uptake per unit time or per unit space), has a negative genetic correlation with life span or survival (e.g., affected by shade tolerance) (Henry and Aarssen, 1997, 2001), then a larger, shorter-lived species may be predicted to have the same lifetime fecundity and hence, the same competitive ability as a smaller, longer-lived species (Fig. 8.8). Likewise, if plant size has a positive genetic correlation with reproductive size threshold, then a larger species is predicted to have greater fecundity at older ages or lower densities, but a smaller species will have greater fecundity at younger ages or higher densities (Fig. 8.9). Similarly, if plant size has a negative genetic correlation with reproductive mass per unit vegetative mass (R/V), then a larger plant with lower R/V may have the same fecundity as a smaller plant with higher R/V (Fig. 8.10a). Finally, if plant size has a negative genetic correlation with fecundity per unit seed

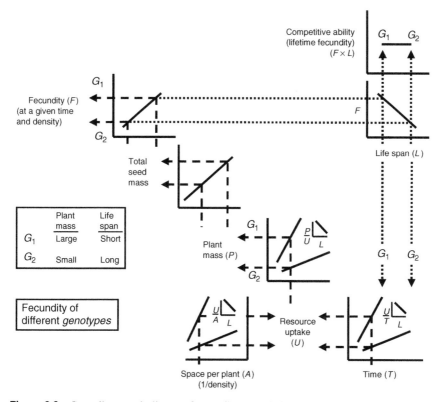

Figure 8.8 Cascading graph diagram for predicting variation in fecundity as a consequence of negatively correlated genotypic variation in plant size and life span. Because plant size or more specifically, plant mass per unit resource uptake (or resource uptake per unit time or per unit space), has a negative genetic correlation with survival or life span (e.g., shade tolerance), then a larger, shorter-lived species (G_1) may be predicted to have the same lifetime fecundity and hence, the same competitive ability as a smaller, longer-lived species (G_2).

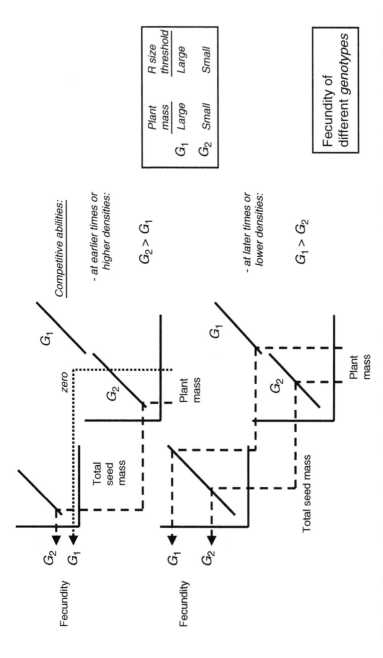

Figure 8.9 Cascading graph diagram for predicting variation in fecundity as a consequence of positively correlated genotypic variation in plant size and reproductive size threshold. A species that is larger but which must also attain a larger size to reproduce (G_1) is predicted to have greater fecundity at older ages or lower densities, whereas a species that is smaller but also able to reproduce at a smaller size (G_2) will have greater fecundity at younger ages or higher densities.

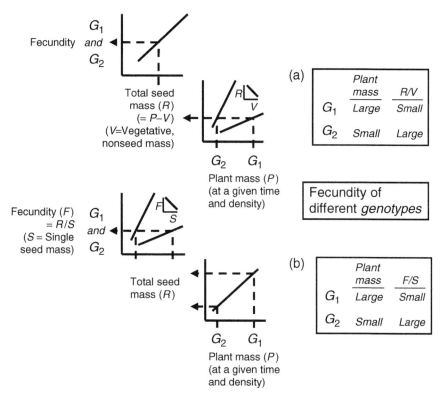

Figure 8.10 Cascading graph diagram for predicting variation in fecundity as a consequence of: (a) negatively correlated genotypic variation in plant size and reproductive mass per unit vegetative mass (R/V); and (b) negatively correlated genotypic variation in plant size and fecundity per unit seed mass (F/S). A larger plant with lower R/V (G_1) may have the same fecundity as a smaller plant with greater R/V (G_2) (a). Similarly, a larger plant with lower F/S (G_1) may have the same fecundity as a smaller plant with greater F/S (G_2) (b).

mass (F/S), then a larger plant with lower F/S may have the same fecundity as a smaller plant with higher F/S (Fig. 8.10b).

V. Preliminary Empirical Tests

Based on these considerations, we should expect that variation in plant size within competing multispecies vegetation will be a poor predictor of variation in fecundity, unless we can also distinguish and account for the effects of variation in species identity (Aarssen and Taylor, 1990), variation in genotype (Aarssen and Clauss, 1992), variation in age/developmental stage (Clauss and Aarssen, 1994a), local variation in the relative importance of different environmental variables (e.g., density, soil fertility, light intensity), that may be causing the variation in plant size (Clauss and Aarssen, 1994b), as well as variation in the degree to which all of these variables interact. Unfortunately, data from competing plants that might be used to test the earlier predictions directly are thus far unavailable. Evidence for some of the aforesaid negative genetic correlations between traits is evident, however, at the species level from a comparative study of 15 monocarpic herbs growing in monospecific natural populations (Aarssen and Jordan, 2001). Vegetative mass at maturity was positively related to both total seed mass (Fig. 8.11a) and fecundity (Fig. 8.11b). However, the most fecund species was not the largest species and the least fecund species was not the smallest species (Fig. 8.11b). Plants of similar size (e.g., the two largest species or the two smallest species, connected by dashed lines in Fig. 8.11) varied widely in fecundity per unit seed mass (by over two orders of magnitude; Fig. 8.11c) having either relatively high fecundity with relatively small seeds or relatively low fecundity with relatively large seeds (Fig. 8.11d). The trade-off (negative relationship) between fecundity and seed size (cf. Fig. 8.10b) was highly significant ($P < 0.0001$) (and proportionate) when covariation in plant size was also accounted for in a multiple regression model ($r^2 = 0.88$) (Aarssen and Jordan, 2001).

Predicting superiority in plant competition becomes even more complicated (and perhaps intractable) when one recognizes that all of the genetic correlations between traits represented in Figs 8.8–8.10 could be important simultaneously in describing patterns of genotypic variation, especially at the scale that spans across taxonomic boundaries within multispecies vegetation. Implications of these potential effects are suggested by data from a series of recent studies involving a group of 10 genotypes (homozygous inbred lines) of *Arabidopsis thaliana* grown singly or in monocultures. Significant genotypic variation has been documented for plant mass per unit nutrient uptake, nutrient uptake per unit time, and nutrient uptake per unit space (Krannitz *et al.*, 1991a), seedling survival under nutrient-starved conditions (Krannitz *et al.*, 1991b), reproductive size threshold (Clauss and Aarssen, 1994b), seed mass (Krannitz *et al.*, 1991b), and plant mass, total seed mass and fecundity at full maturity (Aarssen and Clauss, 1992). Most of these traits appear to be inherited independently as there are very few correlations between these traits across the 10 genotypes. Seed mass is positively correlated with seedling survival under nutrient-deprived conditions (Krannitz *et al.*, 1991b). As for the between-species comparison of monocarpic herbs mentioned earlier, there is also a negative relationship between fecundity and seed size when covariation in plant size is included in multiple regression analysis (Fig. 8.12; see also Figs 8.14c and d). However, whereas the plant size effect on fecundity is positive in the between-species analysis (Fig. 8.11), its effect is negative in the comparison of *Arabidopsis* genotypes (Fig. 8.12); variation in both fecundity (F) and total seed mass (R) is negatively related to (i.e., they "trade-off" with) variation

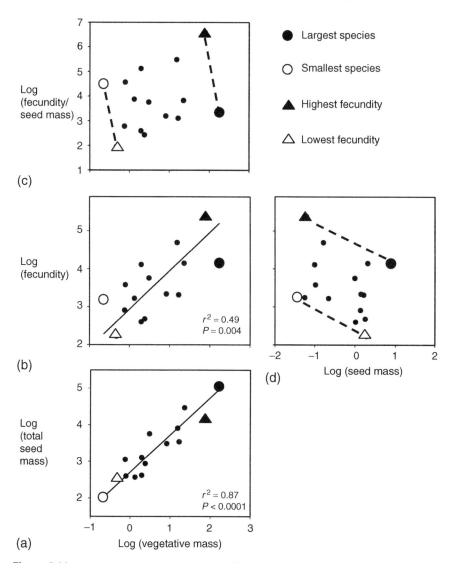

Figure 8.11 Relationships between traits for 15 monocarpic herbs at final developmental stage (from Aarssen and Jordan, 2001). The two smallest species are indicated by open symbols and the two largest species are indicated by shaded symbols. The smallest and largest species are indicated by the open and shaded circles, respectively, whereas the least and most fecund species are indicated by open and shaded triangles, respectively. Dashed lines in (c) and (d) connect species of similar size. Solid lines with r^2 and associated P-values are from reduced major axis (model II) regression analysis (see text).

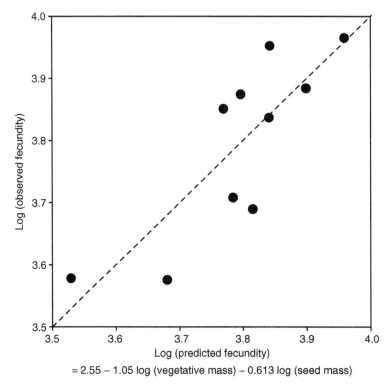

= 2.55 − 1.05 log (vegetative mass) − 0.613 log (seed mass)

Figure 8.12   Log (observed fecundity) versus log (predicted fecundity) from a least squares multiple regression model ($r^2 = 0.68$) involving log (above ground vegetative mass) and log (individual seed mass) for 10 genotypes (inbred lines) of *Arabidposis thaliana* at final developmental stage (full maturity). The allometric equation is given below the axis for log (predicted fecundity); the scaling coefficients are negative for both vegetative mass ($b = −1.05$; $P = 0.007$) and individual seed mass ($b = − 0.613$; $P = 0.061$). (Fecundity and vegetative mass data are from Aarssen and Clauss (1992); seed mass data are from Krannitz *et al.* (1991b).)

in vegetative mass (V) (Fig. 8.13), as depicted by the model in Fig. 8.10. Note that variation in plant size here is entirely due to genotypic variation; plants were all at the same age/developmental stage and all experienced the same environmental conditions.

These trait relationships suggest that it would be implausible to make any predictions of relative competitive abilities for these 10 genotypes based on only a single trait. The rank order of genotyes is highly trait dependent with no significant concordance (Fig. 8.14). The rank order based on plant size alone (Fig. 8.14a) bears no resemblance to the mean rank order (Fig. 8.14e) derived from the mean of ranks for plant size, fecundity (Fig. 8.14b), seed mass (Fig. 8.14c), and seedling longevity under nutrient-deprived conditions (Fig. 8.14d). Moreover, genotypes do not differ

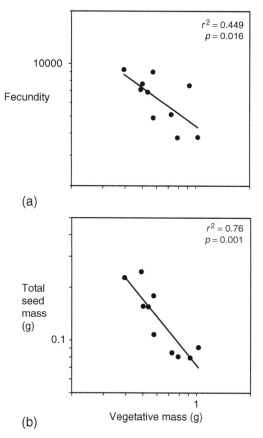

Figure 8.13   Fecundity versus vegetative mass (a) and total seed mass versus vegetative mass (b) for 10 genotypes (inbred lines) of *Arabidposis thaliana* at final developmental stage (full maturity) and grown under identical environmental conditions (note log scale). (Data from Aarssen and Clauss (1992)). Solid lines with r^2 and associated *P*-values are from reduced major axis (model II) regression analysis.

significantly from each other in their mean rank based on all four component trait ranks (Fig. 8.14e, ANOVA, $F = 0.45$, $P = 0.897$).

VI. Predicting Winners from Rank Orders in Plant Competition: Lessons from Sports Competition

Several lines of evidence, from both theoretical and empirical research suggest that the above genetic correlations (Figs 8.8–8.11, 8.13) are likely to be common in plants. If this is true, then we should expect there to be several different ways for a plant to be a good competitor; i.e., there should

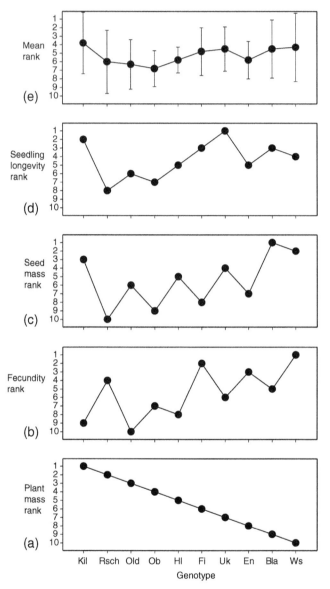

Figure 8.14   Ten genotypes (inbred lines) of *Arabidposis thaliana* ranked in decreasing order of relative plant mass at full maturity (a), with corresponding relative ranks for fecundity at full maturity (b), individual seed mass (c), seedling longevity under nutrient-impoverished conditions (d), and the mean (±SE) of ranks in a–d. (e). Note that a change in relative rank based on fecundity (b) is always accompanied by a rank change in the opposite direction based on seed mass (c); i.e., fecundity and seed mass are negatively correlated for plants of similar size (Fig. 8.12). (Plant mass and fecundity data from Aarssen and Clauss (1992); seed mass and seedling longevity data from Krannitz *et al.* (1991b).)

be several potential combinations of genotypically variable plant traits that are all equally effective in enabling a species or genotype to experience intense competition, yet overcome the risk of exclusion across generations. Extending this prediction to the community scale, these alternative trait combinations should allow several species to compete intensely *together*, while they all avoid exclusion (i.e., coexist). A rigorous test of these predictions would require data that incorporate all of the major components of competitive ability in plants (Fig. 8.2) for a large number of coexisting species within natural vegetation. Such data, however, have apparently never been published (Aarssen and Keogh, 2002). Data are available, however, that incorporate all of the major components of competitive ability for large numbers of competitors in professional sports and where interesting patterns have implications for the interpretation of relative competitive ability and species coexistence in plant communities. The end of the regular season standings for sports teams is represented by a linear rank order with the top rank occupied by the team that accumulated the most points through game wins and ties against all other teams in the league. In North America, professional basketball teams generally differ widely in competitive ability with the top-ranked team earning about 74% wins, on average, and the bottom-ranked team earning only about 24% wins (Fig. 8.15a). The rank order for professional baseball teams, however, looks very different; overall competitive abilities are much more similar with the top- and bottom-ranked teams earning about 61 and 42% wins, on average, respectively (Fig. 8.15b).

Why is this so? Possibly, the variation in team payroll budgets is generally greater in basketball than in baseball, thus causing greater variation between basketball teams in their abilities to afford to attract the most talented players. Alternatively, perhaps the longer season in baseball (about twice that of basketball) allows more time for the less talented teams to improve their relative competitive abilities over the season. A more interesting speculation, however, is based on the proposition that basketball and baseball differ greatly in the number of individual traits that define a team's competitive ability. For basketball, one could argue that the competitive ability of a team is defined largely by a single trait: the ability to maximize the number of baskets scored within the frantic scramble of a fixed game time. To be sure, there is also a defensive component to game strategy, but basketball is probably the best example of where, as the popular sports expression goes, "the best defense is a strong offense." This is reflected, for example, in the final scores of basketball games, which are typically an order of magnitude higher than in baseball.

Compared with basketball, the competitive ability of a baseball team has arguably more numerous and distinctly different offensive and defensive components involving pitching, hitting (batting), running, in-fielding, and out-fielding, along with several corresponding categories of skills

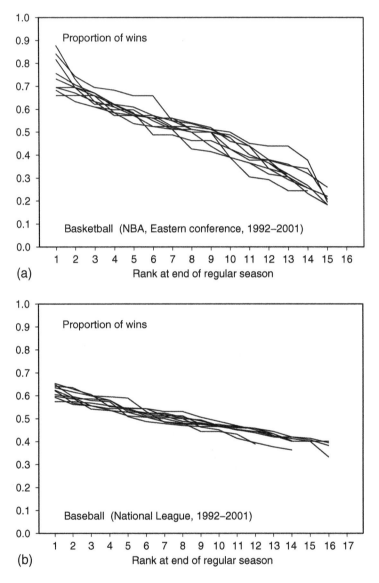

Figure 8.15   End-of-season team rank orders based on proportion of wins in each year from 1992–2001 for (a) Eastern Conference of the National Basketball Association; and (b) Major league baseball (National League). (Data were obtained from league websites and individual team websites on the internet.)

associated with various trainers and coaches. One could of course argue that there are also several different ways that a basketball team might maximize the number of baskets scored; e.g., effective penetration by point guards, accurate outside shooting, or domination by a big man on the inside. In baseball, however, the same level of variety can be found in just pitching strategy alone; e.g., a pitcher may be good because of an effective curve ball, a deceptive drop ball, or a precisely targeted fast ball. Since baseball teams might commonly be expected to have a limited budget, it may only occasionally be possible for a single team to afford to attract better pitchers *plus* betters hitters, *plus* better runners, *plus* better fielders *plus* better trainers *plus* better coaches than every other team in the league. Hence, the top ranking team, although superior to the greatest *number* of other competitors, may not always be superior to *all* of its competitors, especially other highly ranked competitors. The outcomes of pairwise competitions, therefore, should be less predictable and thus, rank orders, less distinct in baseball than in basketball (Fig. 8.15b).

Consider hypothetically, for example, the top three teams ranked $A > B > C$ in terms of their total end-of-season points. This reflects their overall competitive abilities relative to all other teams in the league, but it need not reflect their competitive abilities relative to each other in particular pairwise competitions. A simple example illustrates how this may be so. Assume that in terms of pitching talent, the three teams are ranked $A > B > C$ with only the widest difference, $A > C$ being large enough to affect (i.e. with statistical significance) the outcome of a game. Assume also that in terms of hitting talent, the teams are ranked $C > A > B$ with only the widest difference, $C > B$ being large enough to significantly affect the outcome of a game. Now assume that in terms of fielding talent, $B > C > A$ with again, only the widest difference $B > A$ being large enough to significantly affect the outcome of a game. Hence, when pairwise comparisons of pitching, hitting, and fielding talents are all accounted for, we should expect the competitive abilities of the three teams to be intransitive: i.e., A usually wins against C (because of superior pitching) and C usually wins against B (because of superior hitting), but B usually wins against A (because of superior fielding). Numerous examples of such intransitive loops among groups of teams are evident from the end-of-season statistics of game outcomes for any given year. In playoff competition, therefore, one might expect that the champion team in baseball is not always the team that was ranked first in the overall standings at the end of the regular season. This is indeed the case in the National League where, in the seven years prior to 2002, the league championship went to the first-ranked team only three times; the second- and fourth-ranked teams won the championship once each and the third-ranked team won the championship on two occasions. This contrasts with the National Basketball Association, where over the same time period, the playoff championship (Eastern Conference) was won by the first-ranked team on five occasions and by the second-ranked team on the remaining two.

It is not possible, of course, to be certain that the differing rank order structure in baseball versus basketball competition (Fig. 8.15) can be accounted for by the above proposition that the competitive abilities of baseball teams are determined by a more complex array of distinct attributes. Nevertheless, the fact that this mechanism represents at least a plausible explanation leads to an intriguing thought experiment for plant competition. On this premise, I suggest that plant competition is likely to be more similar to competition in baseball than in basketball. If plant competitive ability were determined by only a single trait such as relative plant size, then rank orders of competitive abilities would be distinct and highly predictable, as in basketball competition. As for baseball teams, however, competitive ability in plants, as proposed earlier (Figs 8.2, 8.8–8.10) is defined by a more complex interaction of several traits. Moreover, because of inherent allocation trade-offs between growth, reproduction and survival, it is not possible for any plant to maximize allocation to all three traits. Hence, when pairwise comparisons of growth under competition, survival under competition, and reproduction under competition are all accounted for within natural vegetation (i.e., across the time scale that encompasses at least the generation time of the longest-lived plant), the competitive abilities (i.e., lifetime offspring production) of three plants may also be intransitive in pairwise (e.g., local neighborhood) contests: e.g., $A>C$ (because of greater plant size and hence ability to deny resources to competitors), $C>B$ (because of greater ability to tolerate or survive resource depletion by competitors), but $B>A$ (because of greater fecundity per unit plant mass and/or per unit time when either is limited by competition). Moreover, because: (i) each species within a community might be expected to have genotypic variation affecting each of these traits; (ii) the genotypic composition of local neighborhoods within vegetation might be expected to be largely stochastically determined; and (iii) because competition in plants is spatially explicit (i.e., primarily between near neighbors only), then the taxonomic identity of A, B, and C in the aforesaid pairwise contests may involve different possible combinations of species in different local neighborhoods. Accordingly, the species identity of the superior competitor would be highly unpredictable at the scale of the local neighborhood. At the scale of the whole community, therefore, the probability of competitive exclusion for any given species would be very low, i.e., species would coexist (Aarssen, 1983, 1989, 1992; Taylor and Aarssen, 1990; see also Frean and Abraham, 2001; Kerr *et al.*, 2002). In this context, "randomness" or "null communities" *sensu* Bell (2001) and Hubbell (2001), involving the coexistence of apparently "equivalent" species at the community scale, is a product not of "chance," but rather a collective product of very deterministic outcome of species interactions occurring within local neighborhoods (Aarssen, 1983). This falls within the general coexistence mechanism that involves small differences in competitive abilities (Fig. 8.1c), but with a broader definition of competitive ability (involving more than just relative plant size), and a different mechanism for maintaining small differences than depicted in Fig. 8.1c.

VII. Conclusions

Plant currencies that impact on fitness (e.g., sequestered nutrients, biomass, meristems) may be allocated to growth, survival, or reproductive functions. The three-way trade-off implied by this allocation defines three corresponding fundamental components of competitive ability in plants (Fig. 8.2) and hence, the potential for intransitive relative competitive abilities of genotypes at the local neighborhood level, thus promoting multispecies coexistence at the community scale, despite intense competition and even within a relatively homogeneous physical environment. Identifying the relevant spatial scales for these patterns and processes, therefore, is of particular importance, but so too is the identification of relevant *temporal* scales. Measurement and comparisons of competitive abilities of plants are incomplete and likely to be misleading unless they encompass both survival and fecundity allocation components spanning at least one complete generation (Aarssen and Keogh, 2002). In addition, species coexistence involving the aforesaid mechanism is not expected to be as theoretically stable as when the minority species always has an advantage (e.g., as with traditional niche differentiation) (Tokeshi, 1999). Theoretically, a "random-walk" to exclusion could be expected eventually for some species that coexist only because of intransitive relative competitive abilities. Nevertheless, the time required for this within real vegetation may be so long that species are more likely to be excluded from a local community for other reasons first (e.g., habitat destruction or climate change). During this time, therefore, species unequivocally coexist (i.e., avoid exclusion despite intense competition), regardless of the fact that there is no theoretical potential for indefinite stable equilibrium (Aarssen, 1983, 1989, 1992; Hubbell and Foster, 1986; Hubbell, 2001). In terms of practical applications (e.g., conservation of biodiversity), the time scales over which it is most relevant to interpret and understand the mechanisms that promote species coexistence and hence, maintain biodiversity within vegetation, are the finite time intervals over which *real* habitats and *real* resident species within them, actually exist.

The challenge for future studies is to obtain quantitative measures of competitive ability for plants in terms of all three fundamental fitness components (Fig. 8.2) and to use these data to test predictions about not only which genotypes and species have the traits necessary to cause the competitive exclusion of others, but also which genotypes and species have the traits necessary to prevail under intense competition and hence, coexist with other genotypes and species that are similarly equipped.

Acknowledgments

Stephen Bonser, Troy Day, Jason Pither, Ed Reekie, and Brandon Schamp provided helpful comments on earlier versions of the manuscript. Data reported here are from research that was supported by the Natural Sciences and Engineering Research Council of Canada through a research grant to the author.

References

Aarssen L. W. (1983) Ecological combining ability and competitive combining ability in plants: Toward a general evolutionary theory of coexistence in systems of competition. *American Naturalist* **122**: 707–731.

Aarssen L. W. (1989) Competitive ability and species coexistence: a 'plant's-eye' view. *Oikos* **56**: 386–401.

Aarssen L. W. (1992) Causes and consequences of variation in competitive ability in plant communities. *Journal of Vegetation Science* **3**: 165–174.

Aarssen L. W. and Clauss M. J. (1992) Genotypic variation in fecundity allocation in *Arabidopsis thaliana*. *Journal of Ecology* **80**: 109–114.

Aarssen L. W. and Jordan C. Y. (2001) Patterns of covariation in plant size, seed size and fecundity in monocarpic herbs. *Ecoscience* **8**: 471–477.

Aarssen L. W. and Keogh T. (2002). Conundrums of competitive ability in plants: what to measure? *Oikos* **96**: 531–542.

Aarssen L.W. and Schamp B. (2002) Predicting distributions of species richness and species size in regional floras: applying the species pool hypothesis to the habitat templet model. *Perspectives in plant ecology, evolution and systematics* **5**: 3–12.

Aarssen L. W. and Taylor D. R. (1992) Fecundity allocation in herbaceous plants. *Oikos* **65**: 225–232.

Agren G. I. and Fagerstrom T. (1984) Limiting dissimilarity in plants: randomness prevents exclusion of species with similar competitive abilities. *Oikos* **43**: 369–375.

Baskin C. C. and Baskin J. M. (1998) *Seeds: Ecology, Biogeography and Evolution of Dormancy and Germination*. Academic Press, London.

Bell G. (2001) Neutral macroecology. *Science* **293**: 2413–2418.

Bengtsson J., Fagerstrom T. and Rydin H. (1994) Competition and coexistence in plant communities. *Trends in Ecology and Evolution* **9**: 246–250.

Berendse F. and Aerts R. (1987) Nitrogen-use efficiency: a biologically meaningful definition? *Functional Ecology* **1**: 293–296.

Bosey J. and Aarssen L. W. (1995) The effect of seed orientation on germination in a uniform environment: differential success without genetic or environmental variation. *Journal of Ecology* **83**: 769–773.

Burns R. M. and Honkala B. H. (1990) *Silvics of North America. Volume 2, Hardwoods*. Agriculture Handbook 654, Forest Service, USDA, Washington.

Chapin F. S. (1980) The mineral nutrition of wild plants. *Annual Review of Ecology and Systematics* **11**: 233–260.

Chesson P. L. (1986) Environmental variation and the coexistence of species. In *Community Ecology* (Diamond J. and Case T. J., eds.), pp. 240–256. Harper and Row, New York.

Clauss M. J. and Aarssen L. W. (1994a) Patterns of reproductive effort in *Arabidopsis thaliana*: confounding effects of size and developmental stage. *Ecoscience* 1: 153–159.

Clauss M. J. and Aarssen L. W. (1994b) Phenotypic plasticity of size-fecundity relationships in *Arabidopsis thaliana. Journal of Ecology* 82: 447–455.

Connell J. (1978) Diversity in tropical rainforests and coral reefs. *Science* 199: 1302–1310.

Connolly J., Wayne P. and Bazzaz F. A. (2001) Interspecific competition in plants: How well do current methods answer fundamental questions. *American Naturalist* 157: 107–125.

Delcourt P. A. and Delcourt H. R. (1987) *Longterm Forest Dynamics of the Temperate Zone*. Ecological Studies 63. Springer Verlag, New York.

Ellner S. (1987) Alternative plant life history strategies and coexistence in randomly varying environments. *Vegetatio* 69: 199–208.

Endler J. A. (1986) *Natural Selection in the Wild*. Princeton University Press. Princeton, New Jersey.

Fagerstrom T. (1988) Lotteries in communities of sessile organisms. *Trends in Ecology and Evolution*. 3: 303–306.

Fox J. F. (1977) Alternation and coexistence of tree species. *American Naturalist* 111: 69–89.

Frean M. and Abraham E. R. (2001) Rock-scissors-paper and the survival of the weakest. *Proc R Soc Biol Sci B* 268: 1323–1327.

Freckleton R. P. and Watkinson A. R. (2000) On detecting and measuring competition in spatially structured plant communities. *Ecology Letters* 3: 423–432.

Gaudet C. L. and Keddy P. A. (1988) A comparative approach to predicting competitive ability from plant traits. *Nature* 334: 242–243.

Goldberg D. E. (1996) Competitive ability: definition, contingency and correlated traits. *Philosophical Transactions of the Royal Society of London, B* 351: 1377–1385.

Grace J. B. (1990) On the relationship between plant traits and competitive ability. In *Perspectives on Plant Competition*. (Grace J. B. and Tilman D., eds.) Academic Press, New York.

Grime J. P. (1979) *Plant Strategies and Vegetation Processes*. Wiley, New York.

Grubb P. J. (1977) The maintenance of species-richness in plant communities: the importance of the regeneration niche. *Biological Reviews* 52: 107–145.

Hardin G. (1968) The tragedy of the commons. *Science* 162: 1243–1248.

Hardin G. (1993) *Living Withing Limits: Ecology, Economics and Population Taboos*. Oxford University Press, New York.

Harper J. L. (1969) The role of predation in vegetational diversity. In Diversity and stability in ecological systems. *Brookhaven Symp Biol* 22: 48–62.

Harper J. L. (1977) *Population Biology of Plants*. Academic Press, London.

Henry H. A. L. and Aarssen L. W. (1997) On the relationship between shade tolerance and shade avoidance strategies in woodland plants. *Oikos* 80: 575–582.

Henry H. A. L. and Aarssen L. W. (2001) Inter- and intra-specific relationships between shade tolerance and shade avoidance in temperate trees. *Oikos* 93: 477–487.

Hubbell S. P. (2001) *The Unified Neutral Theory of Biodiversity and Biogeography*. Princeton University Press, Princeton, New Jersey.

Hubbell S. P. and Foster R. B. (1986) Biology, chance and history and the structure of tropical rainforest communities. In *Community Ecology* (Diamond J. and Case T. J., eds.) pp. 314–329. Harper and Row, New York.

Huston M. A. (1979) A general hypothesis of species diversity. *American Naturalist* 113: 81–101.

Janzen D. H. (1970) Herbivores and the number of tree species in tropical forests. *American Naturalist* 104: 501–528.

Keddy P. A. (1989) *Competition*. Chapman and Hall, London.

Keddy P. A. (1990) Competitive hierarchies and centrifugal organization in plant communities. In *Perspectives on Plant Competition* (Grace J. B. and Tilman D., eds.) pp. 265–290, Academic Press, New York.

Kerr B., Riley M. A., Feldman M. W. and Bohannan B. J. M. (2002) Local dispersal promotes biodiversity in a real-life game of rock-paper-scissors. *Nature* **418**: 171–174.

Krannitz P. G., Aarssen L. W. and Lefebvre D. D. (1991a) Relationships between physiological and morphological attributes related to phosphate uptake in 25 genotypes of *Arabidpsis thaliana. Plant and Soil* **133**: 169–175.

Krannitz P. G., Aarssen L. W. and Dow J. M. (1991b) The effect of genetically-based differences in seed size on seedling survival in *Arabidopsis thaliana* (Brassicaceae). *American Journal of Botany* **78**: 446–450.

Nijs I. and Impens I. (2000) Underlying effects of resource use efficiency in diversity–productivity relationships. *Oikos* **91**: 204–208.

Pianka E. R. (1988) *Evolutionary Ecology*, 4th edition. Harper Collins, New York.

Reekie E. G. and Bazazz F. (1987) Reproductive effort in plants. 3. Effect of reproduction on vegetative activity. *American Naturalist* **129**: 907–919.

Risch S. J. and Carroll C. R. (1986) Effects of seed predation by a tropical ant on competition among weeds. *Ecology* **67**: 1319–1327.

Roach D. A. and Wulff R. D. (1987) Maternal effects in plants. *Annual Review of Ecology and Systematics* **18**: 209–235.

Steward F. M. and Levin B. R. (1973) Partitioning of resources and the outcome of interspecific competition: a model and some general considerations. *American Naturalist* **107**: 171–198.

Taylor D. R. and Aarssen L. W. (1990) Complex competitive relationships among genotypes of three perennial grasses: implication for species coexistence. *American Naturalist* **136**: 305–327.

Tilman D. (1982) *Resource Competition and Community Structure.* Princeton University Press, Princeton, New Jersey.

Tilman D. (1984) Plant dominance along an experimental nutrient gradient. *Ecology* **65**: 1445–1453.

Tilman D. (1986) Resource, competition and the dynamics of plant communities. In *Plant Ecology* (Crawley M. J. eds.) pp. 51–65, Blackwell, Oxford.

Tilman D. (1988) *Plant Strategies and the Dynamics and Structure of Plant Communities.* Princeton University Press, Princeton, NJ.

Tokeshi M. (1999) *Species Coexistence: Ecological and Evolutionary Perspectives.* Blackwell, Oxford.

Index

A

acid rain, 80
allocation
    biomass, 2, 6, 9, 95
    carbon, 2, 9
    energy, 2, 6, 8, 9
    somatic, 22–24, 50, 51
allometric
    analysis, 94, 96
    relationship/s, 16, 94–99, 105, 109,
        112, 119, 121, 198, 199
allometry of RA, 95, 96, 99–101, 105,
    109, 112, 113, 118–121
among-population, 66, 81, 112, 133
anthesis, 66, 75, 88, 128

B

between-sex genetic correlation, 142,
    146, 148
between-species, 214, 215, 223, 228

C

C – S – R theory/strategy, 19
C/N ratio, 197–201, 204
capital breeders, 26–29
cleistogamy/ous, 107, 108
coexisting species, 211, 213, 233
competition theory, 211
cosexual, 78, 138, 139
cost function, 205, 206
cost of reproduction, 4, 20–22, 28, 127,
    128, 133–135, 155, 169, 179
    demographic, 2, 22, 26, 29, 30, 126,
        128–133, 168
    direct, 126, 168, 170
    physiological, 2, 21
    sex-specific, 126, 132
    somatic, 5, 23, 25, 26, 28, 170, 171
currency, 5–7, 10, 23, 49, 50, 69,
    129, 215

D

dormant buds, 56–58, 62, 68, 69
drought stress, 77–79, 81, 82, 88, 89
dry mass, 95, 96, 105, 107, 115, 162, 163,
    169–173, 178, 181

E

evapotranspiration, 79, 82, 83
experimental manipulation, 17, 189,
    190, 192

F

floral traits, 76, 80, 82, 86, 87, 89
fraction of nitrogen in leaves (FNL),
    173–177

G

genetic correlation, 67, 115, 141,
    142, 146, 148, 187, 189, 190, 225,
    228, 231
genotypic variation, 63, 105, 223–228,
    230, 236

H

heliotropism, 83, 85. *See also* solar
    tracking

I

income breeders, 23, 26–29
inflorescence, 50, 51, 52, 54, 56, 57,
    62–64, 66, 67, 68, 70, 80, 99, 117,
    128, 130, 134, 186, 195
intense competition, 211, 213–217, 233,
    237, 238
intrafloral temperature, 83–86

L

life history
    pattern, 187, 188
    strategy/ies, 2, 30, 50, 95, 99, 136

life history (*Continued*)
  theory, 10, 11, 13, 120, 186
  trait, 10, 50, 126, 127, 132, 136

M
mean residence time (MRT), 172–177
meristem
  allocation, 22, 51–53, 55–59, 62–67,
    69–71
  axillary, 51, 52, 54, 56, 57, 62
  floral, 54–57, 77
  plasticity of, 52, 64–66, 70
  shoot apical, 51, 53, 56, 57
multiple limitation hypothesis, 6

N
nectar production, 7, 76, 78, 86, 87
nitrogen productivity (NP), 172–177,
    179, 181
nitrogen uptake, 169, 172–174, 176, 178,
    179, 181, 182, 197
nitrogen use efficiency (NUE), 138, 143,
    144, 172–177
  photosynthetic (PNUE), 135, 136, 147,
    173–177
  reproductive (RNUE), 174, 176,
    177, 182

O
ontogeny, 62, 64, 65, 99

P
phenotypic correlation, 21, 187
photoperiod, 55, 155, 156, 158, 160–162,
    164, 169, 173, 188–194, 196
photosynthesis
  ear, 7
  leaf, 7, 26, 78, 135, 195, 207
  reproductive, 25, 26
physiological cost, 3, 78, 80, 89
plant competition, 66, 69, 215, 228,
    231, 236
plant ecology, 2, 17, 50
plant life history/ies, 1, 11, 50, 95, 98,
    109
plant size, 10, 16, 20, 64, 65, 95, 96, 98,
    99, 112, 114, 117, 121, 131, 185,
    190, 198, 202, 204, 206–208, 214,
    216, 217, 220, 222, 223, 225–228,
    230, 236, 237
plant water relation. *See* water relation
plant–pollinator interaction, 86, 88, 89

pollen germination/ability, 79, 80–83
population dynamics, 50, 58, 95
principle of allocation, 1–3, 6, 50, 56,
    99, 168

R
regression
  analysis, 86, 96, 228, 229, 231
  model, 96, 113, 228, 230
  result, 103, 106, 110, 111, 113, 114,
    116, 117
relative fitness, 102–106, 121
relative somatic cost (RSC), 4, 23, 25, 26,
    28, 30, 170–172, 176
reproductive allocation (RA), 1, 2, 4–9,
    12, 22–28, 30, 50, 51, 53, 56, 58–64,
    66, 69, 70, 94, 95, 97, 98, 102, 103,
    106, 107, 110, 113, 116, 118, 120,
    126, 140, 141, 165, 171, 185,
    187–189, 195–197, 207, 223
  age and, 13
  empirical patterns, 9, 10, 14
  genetic variation in, 19
  lifetime, 8
  mass and, 109, 112, 115
  short-term, 8
  size-dependence effects, 10, 13,
    16, 17
  standing, 8
reproductive biomass, 26, 129, 165, 178,
    192, 199, 204–206
reproductive effort (RE), 1–5, 7, 10,
    23, 70, 96, 98, 139, 168, 170, 171,
    176, 177
  definition, 3, 4, 7, 8
  direct/indirect measures of, 4
reproductive function, 100, 127, 140,
    192, 237
reproductive investment, 2–4, 9, 20, 23,
    26, 70, 128, 131–135, 137, 140, 141,
    146, 147, 190, 191, 199, 201, 202,
    204–206
reproductive mass, 96, 97, 99, 100, 169,
    181, 192–194, 196, 198, 200, 203,
    205, 206
reproductive output, 1, 58, 69, 95, 99,
    109, 117, 155, 156, 158, 161, 162,
    164, 165, 169, 172, 182, 186, 187,
    192, 194, 207, 208
reproductive part, 7–9, 21, 22, 26, 29, 50,
    96, 155, 157, 161, 162, 164, 165,
    170, 178, 180

reproductive ratio (RR), 157, 158, 161, 162, 165
resource cost hypothesis, 76, 80, 81
resource economy, 1–3, 11, 22, 23
resource investment, 1, 22, 30, 70, 141
resource use efficiency, 133–135, 138
r–K theory, 2, 10, 14, 19, 20, 50

S
seed mass, 98, 100, 112, 118, 178, 216, 218, 219, 221, 223, 224, 226–232
seed size, 2, 85, 100, 228
senescence of leaves, 164, 168, 169
sex function, 126, 128, 137, 139
 female/male, 79, 128–130, 133, 138, 139
sex morph, 78, 127, 131–133, 136, 137, 139, 141, 142
sex-specific physiology, 126–128, 136, 140–142, 146, 148
sexual dimorphism, 127, 131, 140–142, 148
shady population, 66
solar tracking, 83, 85, 86. *See also* heliotropism
successional habitat, 14, 15, 41, 42, 97
sunny population, 66, 68

T
trade-off function, 185–188, 190–192, 194, 195, 197–202, 204, 206–208
traits
 morphological, 94, 141
 physiological, 87, 89, 119, 120, 127, 128, 136, 139–144, 146
 plant, 119–121, 220, 233
 reproductive, 142,146

V
vegetative biomass, 161, 192, 199, 204–206
vegetative body, 76, 155–157, 162, 164, 165, 171, 178–181
vegetative mass, 95–100, 102, 103, 107, 109–117, 119, 120, 121, 163, 193, 194, 196, 200, 205, 221, 224, 225, 227–231

W
water availability, 75, 81, 86–88
water cost, 75, 78, 80, 138
water relation, 75, 76, 80, 82, 83, 89
water use efficiency (WUE), 82, 135, 138, 141, 143, 147
water use, 75, 76, 78–80, 83, 89
within-species, 215, 223

Physiological Ecology
A Series of Monographs, Texts, and Treatises

T. T. KOZLOWSKI. Growth and Development of Trees, Volumes I and II, 1971

D. HILLEL. Soil and Water: Physical Principles and Processes, 1971

V. B. YOUNGER and C. M. McKELL (Eds.). The Biology and Utilization of Grasses, 1972

J. B. MUDD and T. T. KOZLOWSKI (Eds.). Responses of Plants to Air Pollution, 1975

R. DAUBENMIRE. Plant Geography, 1978

J. LEVITT. Responses of Plants to Environmental Stresses, Second Edition
Volume I: Chilling, Freezing, and High Temperature Stresses, 1980
Volume II: Water, Radiation, Salt, and Other Stresses, 1980

J. A. LARSEN (Ed.). The Boreal Ecosystem, 1980

S. A. GAUTHREAUX, JR. (Ed.). Animal Migration, Orientation, and Navigation, 1981

F. J. VERNBERG and W. B. VERNBERG (Eds.). Functional Adaptations of Marine Organisms, 1981

R. D. DURBIN (Ed.). Toxins in Plant Disease, 1981

C. P. LYMAN, J. S. WILLIS, A. MALAN, and L. C. H. WANG. Hibernation and Torpor in Mammals and Birds, 1982

T. T. KOZLOWSKI (Ed.). Flooding and Plant Growth, 1984

E. L. RICE. Allelopathy, Second Edition, 1984

M. L. CODY (Ed.). Habitat Selection in Birds, 1985

R. J. HAYNES, K. C. CAMERON, K. M. GOH, and R. R. SHERLOCK (Eds.). Mineral Nitrogen in the Plant–Soil System, 1986

T. T. KOZLOWSKI, P. J. KRAMER, and S. G. PALLARDY. The Physiological Ecology of Woody Plants, 1991

H. A. MOONEY, W. E. WINNER, and E. J. PELL (Eds.). Response of Plants to Multiple Stresses, 1991

F. S. CHAPIN III, R. L. JEFFERIES, J. F. REYNOLDS, G. R. SHAVER, and J. SVOBODA (Eds.). Arctic Ecosystems in a Changing Climate: An Ecophysiological Perspective, 1991

T. D. SHARKEY, E. A. HOLLAND, and H. A. MOONEY (Eds.). Trace Gas Emissions by Plants, 1991

U. SEELIGER (Ed.). Coastal Plant Communities of Latin America, 1992

JAMES R. EHLERINGER and CHRISTOPHER B. FIELD (Eds.). Scaling Physiological Processes: Leaf to Globe, 1993

JAMES R. EHLERINGER, ANTHONY E. HALL, and GRAHAM D. FARQUHAR (Eds.). Stable Isotopes and Plant Carbon–Water Relations, 1993

E. -D. SCHULZE (Ed.). Flux Control in Biological Systems, 1993

MARTYN M. CALDWELL and ROBERT W. PEARCY (Eds.). Exploitation of Environmental Heterogeneity by Plants: Ecophysiological Processes Above- and Belowground, 1994

WILLIAM K. SMITH and THOMAS M. HINCKLEY (Eds.). Resource Physiology of Conifers: Acquisition, Allocation, and Utilization, 1995

WILLIAM K. SMITH and THOMAS M. HINCKLEY (Eds.). Ecophysiology of Coniferous Forests, 1995

MARGARET D. LOWMAN and NALINI M. NADKHARNI (Eds.). Forest Canopies, 1995

BARBARA L. GARTNER (Ed.). Plant Stems: Physiology and Functional Morphology, 1995

GEORGE W. KOCH and HAROLD A. MOONEY (Eds.). Carbon Dioxide and Terrestrial Ecosystems, 1996

CHRISTIAN KÖRNER and FAKHRI A. BAZZAZ (Eds.). Carbon Dioxide, Populations, and Communities, 1996

FAKHRI A. BAZZAZ and JOHN GRACE (Eds.). Plant Resource Allocation, 1997

J.J. LANDSBERG and S.T. GOWER. Application of Physiological Ecology to Forest Management, 1997

THEODORE T. KOZLOWSKI and STEPHEN G. PALLARDY (Eds.). Growth Control in Woody Plants, 1997

LOUISE E. JACKSON (Ed.). Ecology in Agriculture, 1997

ROWAN F. SAGE and RUSSELL K. MONSON (Eds.). C_4 Plant Biology, 1999

YIGI LUO and HAROLD MOONEY (Eds.). Carbon Dioxide and Environmental Stress, 1999

JACQUES ROY, BERNARD SAUGIER, and HAROLD A. MOONEY (Eds.). Terrestrial Global Productivity, 2001

L.B. FLANAGAN, J.R. EHLERINGER, and D.E. PATAKI (Eds.). Stable Isotopes and Biosphere-Atmosphere Interactions: Processes and Biological Controls, 2005

N. MICHELE HOLBROOK and MACIEJ A. ZWIENIECKI (Eds.). Vascular Transport in Plants, 2005

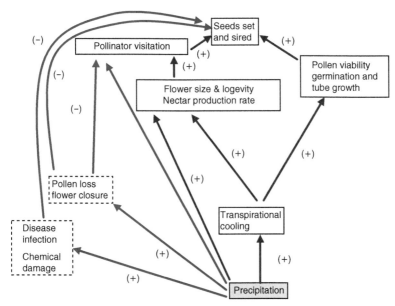

Figure 3.1 Multiple causal pathways linking water input during flowering to plant reproductive success. Red lines and dashed boxes indicate negative effects of water mediated through environmental conditions during flowering. Blue lines and solid boxes indicate positive effects of water as a limiting resource for flower functions.

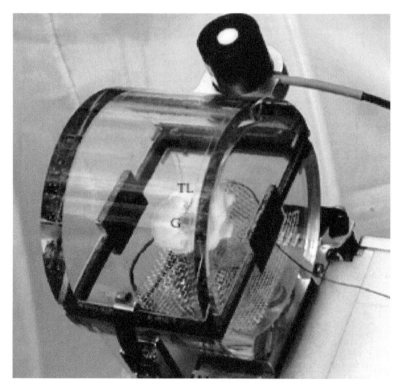

Figure 3.3 Modified "Conifer chamber" (Li-Cor Inc., 6400-05) adapted for measurements of gas exchange and intrafloral temperature in flowers of the snow buttercup, *Ranunculus adoneus*. TL indicates thermocouple lead to gynoecia (G).

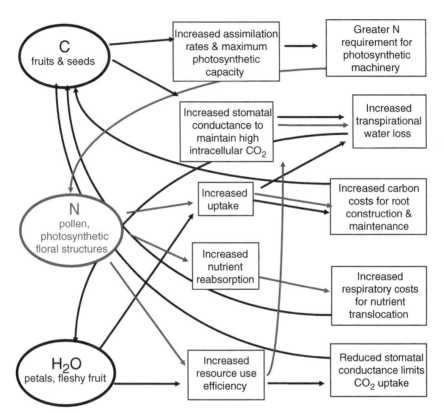

Figure 5.2 Some examples of physiological mechanisms for accommodating proximal costs of reproductive investment by plants. Straight arrows originate from one of three types of resource in greatest demand (left column): C, black = carbon; N, red = nitrogen or other mineral nutrients; H_2O, blue = water. The central column of boxes lists possible strategies for reducing resource limitation. The right column of boxes and the curved arrows indicate physiological consequences of adjustments listed in the central boxes.

(a)

(b)

Figure 6.2 (a) Lake Kamahusa with *Xanthium canadense* Mill. population on the shore, a late summer view. (b) A close-up of *X. canadense* population.

Printed and bound by CPI Group (UK) Ltd, Croydon, CR0 4YY

03/10/2024

01040415-0004